Ultrafast Lasers and Optics for Experimentalists

Ultrafast Lasers and Optics for Experimentalists

James David Pickering

Department of Chemistry, Aarhus University, Aarhus, Denmark

IOP Publishing, Bristol, UK

ISBN   978-0-7503-3659-8 (ebook)
ISBN   978-0-7503-3657-4 (print)
ISBN   978-0-7503-3660-4 (myPrint)
ISBN   978-0-7503-3658-1 (mobi)

DOI   10.1088/978-0-7503-3659-8

Version: 20210301

IOP ebooks

British Library Cataloguing-in-Publication Data: A catalogue record for this book is available from the British Library.

Published by IOP Publishing, wholly owned by The Institute of Physics, London

IOP Publishing, Temple Circus, Temple Way, Bristol, BS1 6HG, UK

US Office: IOP Publishing, Inc., 190 North Independence Mall West, Suite 601, Philadelphia, PA 19106, USA

To my mother, father, and sister, for their unwavering love and support.

Contents

Preface

This book grew as I was preparing to give a short talk to my group (in chemistry) about the basic principles of ultrafast optics and lasers that I think are relevant for non-specialists using 'turn-key' systems. The project rather ran away with me and I ended up producing an extended handout as a supplement to the slides, which eventually turned into this book. My intention is that this will be useful as an introductory text and basic reference for anyone new to the exciting world of ultrafast optics and lasers—and is specifically aimed towards those entering the field without an extensive formal background in optics or laser physics (such as myself). As somebody who initially trained as a chemist, but that moved into a group which was much more physics-based for a PhD, there was a serious learning curve to becoming competent in the use of ultrafast laser systems that the research work demanded. To my knowledge, there are minimal written resources codifying the basic principles of practical use of ultrafast lasers and optics—with most knowledge being passed on orally from postdocs to students. This book aims to fill that gap, and is written largely to be 'the book I wish I had when starting my PhD'.

The aim therefore is to present the key information about practical use of ultrafast lasers and optics in a way that is accessible to non-specialists working in spectroscopy (or related fields) who don't have the formal background in optics, electromagnetism, or laser physics that would be gained in a typical undergraduate physics course. Minimal prior knowledge is assumed (only that which may have been gained in typical undergraduate spectroscopy or quantum mechanics courses), and I do not believe that for this purpose the material can be presented too simply. Emphasis throughout is placed on qualitative, intuitive, understanding rather than dense mathematics. I hope that it will allow students and researchers using these systems to not see lasers as a mysterious 'black box' but rather begin to appreciate the underlying physics. As more and more research groups across the natural sciences are starting to use ultrafast laser systems in their research, I hope that this book can prove to be a useful companion for people moving into the field, both young and old.

The structure of the book is divided into two parts. The first part, '*Fundamentals*', is written with the aim of bringing a final year undergraduate student unfamiliar with lasers quickly up to speed with accessible descriptions of the most important phenomena surrounding ultrafast laser physics. The basic principles of laser action will be discussed, and the properties of laser light and laser beams—with a special focus on those properties that especially differentiate an *ultrafast* laser from a non-ultrafast laser. Useful phenomena within non-linear optics will be covered, focussing on those phenomena which are critical to a working understanding of ultrafast pulse generation and characterisation. These latter two points (generation and character-isation) are then dealt with in some detail—including information on beam characterisation that is not specific to ultrafast lasers, but is essential for a student to know nonetheless. The second part of the book, '*Practical Ultrafast Optics*', essentially aims to walk the student through the construction of an ultrafast beamline. Things that perhaps seem mundane to a more experienced campaigner, but can often seem bewildering to a new student (such as what to look for when buying optics), are discussed in detail. Techniques for practical alignment of optical beamlines are written down, and finally an illustrative example of a beamline for an ultrafast pump-probe experiment is given. The examples given throughout this book all focus on systems driven by Titanium Sapphire lasers, but most of the discussion applies to other types of ultrafast lasers too (such as the increasingly common Yb:Fibre laser).

It is my hope that the book can be used not only by new students, but also by more experienced scientists who are moving into this field. For this reason, it has been written (as far as possible) to be easily dipped in and out of—so that an experienced postdoc does not need to read lots of information on basic laser action in order to follow the narrative. Hopefully this book can form a useful reference text for people in many laboratories. Suggestions for further reading are given throughout, as this book is only intended to provide an accessible gateway into this vast topic.

Common laser terminology

You will encounter the following terms throughout the book (and during discussions with colleagues). They are defined here as used in the book for quick reference.

- **Frequency**, ω: The number of oscillations per unit time in a light field. Strictly we use angular frequency, but it is commonly referred to as simply 'frequency'.
- **Wavelength**, λ: The distance travelled by the light field in one oscillation.
- **Wavenumber**, k: The number of oscillations per unit length in a light field. The wavenumber is inversely proportional to wavelength, and directly proportional to energy.
- **Bandwidth**, $\Delta\omega$: The range of frequencies present in a pulse, with units of frequency or wavelength. The FHWM width of the pulse in the frequency domain.
- **Speed of Light**, c_0: The speed of light in a vacuum. 3×10^8 m s^{-1}.
- **Refractive Index**, n: A factor which determines the speed of light in a medium other than vacuum.
- **Phase Velocity**, v_{p}: The speed of a light wave in a given medium. Related to the speed of light in a vacuum by c_0/n.
- **Spectral Phase**, $\phi(\omega)$: The phase ϕ of a particular frequency ω in a laser pulse.
- **Pulse Duration**, $\Delta\tau$: The FWHM width of the pulse envelope in the time domain.
- **Chirp**: The time-dependent frequency of a pulse. If the frequency changes during the pulse it is 'chirped'.
- **Chromatic Dispersion**: The spreading out of frequency components in time as a pulse propagates. Also called 'temporal dispersion' or simply 'dispersion'.
- **Propagation direction**: The direction of travel of a laser beam, defined by the direction of the wave vector, $\boldsymbol{k}$. The wavenumber is the magnitude of the wave vector.
- **Polarisation direction**: The direction of the oscillations of the electric field of the laser beam. The polarisation plane is orthogonal to the propagation direction.
- **Beam waist**, x_0: Defined as the radius from the centre of the beam at which the intensity is reduced to a factor of $1/e^2$ of the peak intensity.
- **Rayleigh Length**, z_R: Defined as the distance along the propagation direction at which the beam waist is $\sqrt{2}$ times larger than at its smallest point.
- **Repetition rate**: How many pulses per second a laser produces.
- **Pulse energy**: The total energy contained within each pulse.
- **Average Power**: The average power transmitted in the laser beam.
- **Peak Power**: The maximum power transmitted in a single laser pulse.
- **Fluence**: The energy per unit area of a given laser pulse.
- **Intensity**: The power per unit area of a given laser pulse.

Foreword

In the 1960s, the same decade as the invention of the laser, the writer Frank Herbert lamented that the 'scientists were wrong, that the most persistent principles of the Universe were accident and error'. Anyone labouring in an academic laboratory, strained by the fast rate of staff and student turnover, will eventually encounter the consequences of this adage. This will be truer still when the research in question involves the intersection of several scientific fields, where the traditional divisions used to demarcate undergraduate degrees break down. As such, any postgraduate student conducting research in one of these areas, as the author James Pickering did when he undertook a PhD in chemical physics, will often find that resources are aimed at the expert rather than the student. Researchers involved with laser applications are notably vulnerable. Lasers are used in a wide variety of techniques in modern experimental chemistry and physics, particularly due to their intersection with analytical methods such as mass spectrometry, but they form only a small part of undergraduate curricula and younger students often only have limited access to them due to the risks involved. This is a real detriment that only serves to compound the potential for 'accident and error' raised by Herbert.

This text counters the above problem by focussing on the practical aspects of ultrafast laser use, and is written as a handbook that students can refer to as needed. The concepts are clearly explained at a level that will be comfortable for upper-year undergraduates, and are written in a way that will help provide a common language that students can use to accurately and confidently present their work to a broader scientific audience. Any scientist delving into ultrafast laser applications will find something to learn from this book. The first section covers the nature and generation of ultrafast laser pulses, their dispersion through media, and techniques to measure them. The second describes how to prepare an instrument beamline for ultrafast pulses, including pragmatic descriptions of day-to-day considerations such as purchasing optics and optomechanics, over-lapping and focussing beams, and outlining a laser table arrangement. The latter aspect is a major achievement of this book: case studies of laboratories are rare in literature, but are invaluable in helping young scientists to make a start while avoiding mistakes.

James Pickering is a talented scientist with many years of experience in research and instrument development. He refined his teaching ability while working as a teaching fellow at the University of Leicester and as a tutorial lecturer at the University of Oxford. Throughout this, he has always demonstrated an ability to straightforwardly explain complex concepts to students in an approachable way. I have been lucky to work with James at various stages of his career, including his master's degree at the University of Oxford, PhD at Aarhus University, and as his postdoctoral supervisor during his return to Oxford in 2019. He is always eager to talk about the practical aspects of science, and this aspect of his nature is evident in this book. James was responsible for developing the beamline in my own laboratory. The 'accident and error' we shared during those days motivated this text, and I believe the reader will benefit from that experience.

—Michael Burt, Oxford, 2020.

Acknowledgements

Whilst the cover shows a single author, in reality this book could not have been written without the knowledge and experience gained from years of discussion and thousands of hours spent in the lab with countless colleagues, supervisors, and friends. However, there are individuals that have made a significant contribution to the production of this text, whom I would like to mention specifically.

The first thanks must go to Claire Vallance (University of Oxford), as without her encouraging words and putting me in touch with IOP, I would not have considered turning the initial lecture handout into this book. In a similar vein, Michael Burt (University of Oxford) was always encouraging, and many of the pitfalls we encountered together are recounted here so they can be avoided by future generations. Both Claire and Michael also helped to proofread parts of this manuscript. Ensuring a manuscript like this is as error-free as possible is a challenge, and I am especially grateful for the expert proofreading and advice provided by Adam Chatterley, Constant Schouder, and Jan Thøgersen (Aarhus University). In addition, some of my more diligent students from Merton College gave invaluable advice on explanations that were unclear or confusing, and the advice of Beth, Nana, Jia Jie, and Frank was particularly valued. Any errors that remain after such expert guidance are, of course, my own.

Finally, any teacher is aware that the only way to truly learn and understand a topic is to teach it to others. I have been fortunate enough to work with and teach many exceptional students over the years, and much of the time I learn more from them than I feel they do from me. I am indebted to them all.

Author biography

James David Pickering

James David Pickering is an experimental physical chemist currently working as a postdoctoral research associate at Aarhus University. Originally from Essex, he attended Notley High School and Braintree Sixth Form, and obtained his MChem in Chemistry at Jesus College, University of Oxford, and his PhD in Chemistry at Aarhus University. Following this he returned to the UK and worked as a postdoctoral researcher at the University of Oxford, where he also taught extensively in physical chemistry and mathematics. His research interests lie in the application of ultrafast laser spectroscopy to real-world chemical problems.

James is a committed and passionate scientific educator and teaches extensively across the physical natural sciences. Most recently, he has taught physical chemistry and mathematics to undergraduates in teaching lectureships at the University of Oxford; and has previously worked as a teaching fellow at the University of Leicester. He is an associate fellow of the Higher Education Academy.

http://jamesdpickering.com/

Author Biography

Part I

Fundamentals

IOP Publishing

Ultrafast Lasers and Optics for Experimentalists

James David Pickering

Chapter 1

Lasers

'LASER' is an acronym, standing for **L**ight **A**mplification by **S**timulated **E**mission of **R**adiation. Lasers are one of, if not the, most widely used experimental tools across much of physical chemistry/chemical physics. Fundamentally, a laser is a device that produces highly directional[1] electromagnetic (EM) radiation (light). This radiation can have a very well-defined energy (with a narrow *bandwidth*, termed '**narrowband** radiation'), which is useful for many kinds of resonant spectroscopy. Alternatively, it can exist as very fast pulses of energy, which necessarily have a poorly-defined energy (a broad bandwidth, or '**broadband** radiation'). Such pulses are very useful for time-resolved spectroscopy of ultrafast processes.

1.1 Why lasers?

The utility of lasers for experimental scientists lies in the aforementioned directionality of the emitted light, and in the relative ease with which a wide variety of different wavelengths of EM radiation can be produced. Radiation that is resonant with a wide range of different molecular energy level spacings can be produced, and as such lasers have become a standard weapon in the armoury of the spectroscopist. The high directionality means that almost all of the produced photons can be efficiently directed onto the sample of interest, so there are not a large number of 'wasted' photons, which is the case when a sample is irradiated using (for example) a discharge lamp.

Some types of lasers emit light (or 'lase') at a single fixed frequency, but others can be very widely tunable to allow a wide range of molecular resonances to be interrogated using a single laser system. Acquiring a laser system that will emit at your desired frequency (or range of frequencies) is (mostly) a matter of finances. Though some frequency ranges are more challenging than others—for example,

---

[1] We say the light has high *Spatial Coherence*.

1-1

producing powerful lasers operating in the X-ray region is difficult in a normal 'tabletop' laboratory laser setup.

1.2 Laser action

Here we will discuss some of the basic principles of laser action—how lasers work. As the acronym suggests, lasers work via stimulated emission of radiation. Figure 1.1 shows a basic schematic of how a four-level laser functions. Other energy level structures are possible (such as a 'three-level laser'), depending on the type of laser, but the basic principles of laser operation are common across all laser types. Here we choose to focus on the four-level laser as this is how a Titanium Sapphire laser functions, which is the type of laser we will focus on for much of the discussion in the remainder of this book.

There are four distinct energy levels in the system shown in figure 1.1. A ground level (1), is excited into level 4, often via optical excitation with photons of energy $\hbar\omega_{pump}$ provided by a flashlamp or another laser (known as a **pump laser**). Level 4 is shown as a broad band of levels, rather than a single level, and this is often the case

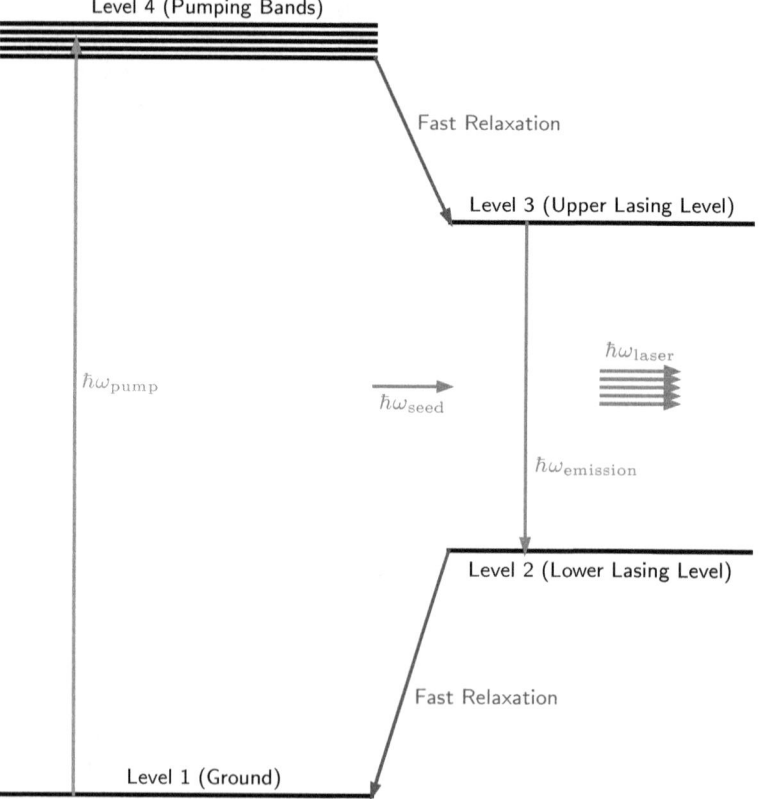

Figure 1.1. A schematic of basic laser action. Lasing is achieved between levels 2 and 3, via excitation from level 1 to level 4.

in real systems. Precise frequency matching of the pumping radiation to a specific upper level is then not necessary, allowing the use of broadband excitation sources.

There is then rapid population transfer from the excitation level(s) 4, into the upper lasing level, level 3. Lasing occurs on the 3 → 2 transition, and requires a **population inversion** to exist between levels 2 and 3. The idea of a population inversion may be familiar from undergraduate spectroscopy courses, and is the situation in which there is more population in the upper level (3) than the lower level (2). This is sometimes referred to as 'non-Boltzmann behaviour'. Achieving the population inversion requires that the population of level 3 is kept high, while the population of level 2 is kept low. This is achieved by excitation from level 1 to level 4, followed by the rapid relaxation into level 3. When the population of level 3 is increased to give the population inversion, we say that level 3 is being **optically pumped**. Stimulated emission from level 3 into level 2 then produces the laser radiation, with the energy gap between these levels determining the frequency (colour) of the laser radiation. This emission is either stimulated using a **seed laser** with photon energy $\hbar\omega_{seed}$ as shown, or relies on spontaneous emission from level 3 of a nearby system to induce the stimulated emission. Another rapid population transfer from level 2 back to the ground state has the effect of ensuring that the population of the lower lasing level remains low, such that a large population inversion can be maintained for efficient lasing.

Clearly, there must be some material in which all these levels exist, and this is known as the **gain medium** or **laser medium**. There are a huge number of different possible laser media, spanning the gas, solution, and solid phases. Here we will limit our discussion to solid-state laser media, as these are the types most commonly found in modern ultrafast lasers[2]. Different laser media have different emission energies (the gap between levels 2 and 3), and different pump energies (the gap between levels 1 and 4). Solid-state lasers are generally centred around an **active ion**. This ion is 'doped' into a host glass or crystal, and the energy levels of the ion *within this crystal field* are what determine the laser action. Common active ions are Nd^{3+}, Ti^{3+}, and Yb^{3+}, and they can be doped into a variety of hosts to give a range of different laser actions. Examples of lasing schemes that can be produced by these active ions doped into common hosts[3] are shown in table 1.1, note the standard nomenclature of '*ion:host*'. YAG (Yttrium Aluminium Garnet) and YLF (Yttrium Aluminium Fluoride) are common hosts for Nd^{3+} ions. They are both included in table 1.1 to illustrate that subtle changes to the chemical structure of the host can be used to tune the emission energy.

Note that the emission energies in table 1.1 need not correspond to a single well-defined energy, as levels 2 and 3 can just as easily be broader bands that lead to a wider lasing bandwidth—this is in fact necessary for ultrafast laser pulse generation. The pump and emission energies shown are merely selected for illustrative reasons. Titanium Sapphire (Ti:Sa) has the broadest emission bandwidth of any solid-state

---

[2] The earliest femtosecond lasers were 'dye lasers', where the gain medium was a solution of organic dye. These are now relatively rare.

[3] Many laser media can be pumped to lase at a variety of different wavelengths. For example, Nd:YAG can be pumped to lase at 1440 nm, as an alternative to its 'usual' emission at 1064 nm.

Table 1.1. Common solid-state laser media and their characteristic lasing energies.

Laser medium	Pump energy (nm)	Emission energy (nm)
Ti:Sapphire	400–600	650–1100
Nd:YAG	730–820	1064
Nd:YLF	730–820	1053
Yb:fibre	940	1020–1040

laser medium, making it exceptionally well suited for the production of ultrashort pulses. It is often pumped at around 527 nm to produce emission between 770 nm and 830 nm. Nd:YAG and Nd:YLF are commonly used laser media for non-ultrafast, high power lasers; with Nd:YLF often being used as a pump laser for Ti:Sa— frequency doubling the 1053 nm output leads to the 527 nm output previously mentioned. Yb^{3+} ions can be doped into a variety of media (including YAG, to make Yb:YAG lasers). Yb-doped fibre lasers in particular are becoming more and more common in ultrafast laboratories worldwide. We will use the Ti:Sa laser whenever we need an example throughout this book, as it is the most commonly encountered kind of ultrafast laser. A more detailed description of how ultrafast lasers operate is provided in chapter 5.

1.3 Oscillators and amplifiers

The laser media mentioned in the previous section can be set up in such a way as to function either as laser **oscillators** or as laser **amplifiers**. A detailed mathematical discussion of these is beyond the scope of this book, but it is important to know the qualitative difference and to understand the basic principles of operation. More detailed descriptions can be found in references [1–4]. Most lasers that you will use in the laboratory are really better defined as laser *systems*, which consist of multiple different parts that produce the final laser output desired. Generally, a laser system will consist of an oscillator and at least one amplifier.

1.3.1 Amplifiers

A laser **amplifier** is a part of a laser which **amplifies** an existing laser beam up to a desired power level. There are a multitude of different geometries of laser amplifier, and some of these will be discussed in due course, but some features are common to all. In all of the cases we will discuss in the context of ultrafast optics, amplification requires a laser medium, a pump laser[4], and a weaker seed laser. Energy is dumped into the laser medium by the pump laser, and is then 'picked up' by the seed laser beam. The effect of the amplifier is that the energy of the pump beam is reduced, whereas the energy of the seed beam is increased. This is exactly the situation drawn in figure 1.1, where $\hbar\omega_{seed}$ is amplified. Figure 1.2 shows a simplified schematic of the

---

[4] Some non-ultrafast amplifiers can use a flashlamp as a pump radiation source, such as Nd:YAG lasers.

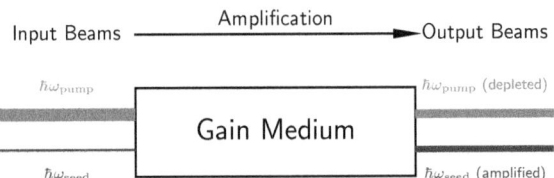

Figure 1.2. Simplified schematic of how a laser amplifier works. The incoming seed beam (red) is amplified, and the pump beam (green) is depleted, via interaction in the gain medium.

amplification process for illustrative purposes. Note that in figure 1.2 the laser medium is referred to as a **gain medium**—the two terms are entirely interchangeable, but 'gain medium' is more common terminology when referring specifically to amplifiers.

The extent to which the energy of the seed beam is increased is referred to as the **gain** of the amplifier, and is often defined as the ratio of the output power to the input power. As may be expected from the preceding discussion in this chapter, laser amplifiers can generally only amplify across a certain range of frequencies, as the gain medium has a finite emission bandwidth. The bandwidth where a laser amplifier can effectively amplify is known as the **gain bandwidth**. For ultrafast laser systems, we desire a *broad* gain bandwidth, as we have to make sure we effectively amplify the full bandwidth of our ultrashort pulse. When the amplifier cannot amplify the full bandwidth of the input pulse, it leads to a situation called **gain narrowing**—the reduction in bandwidth of a pulse following amplification. Often there are multiple amplification stages within a laser system. These can broadly be divided into **preamplifiers** and **power amplifiers**, depending on the gain produced[5].

1.3.2 Oscillators

A laser **oscillator** is a part of a laser which *generates the initial, weak, laser pulses*. These weak pulses are normally then used as a *seed* for an amplifier later on inside the laser system. In a simple case, taking a laser amplifier and applying feedback to it (i.e. returning some of the amplified output to the input of the amplifier) will cause it to **oscillate** at the frequency of the radiation that is fed back. Simplistically this can be thought of as taking the output light from a stimulated emission process, and re-routing it back into the input, inducing more of the same kind of stimulated emission. Oscillation is practically achieved by building an **optical cavity** around an amplifier, so that the laser beam resonates around the cavity, passing through the gain medium on each round trip, then the emission is returned to the input by the cavity, stimulating more emission, and oscillation occurs. An optical cavity is simply a region enclosed by mirrors, such that a laser beam injected into it will bounce round and round it rather than passing straight through. A simplified schematic of how an amplifier can be turned into an oscillator is shown in figure 1.3.

---

[5] The preamp/power amp definition is less well defined than in electronic amplifiers, where preamplifiers typically provide high voltage gain but relatively little current gain, whereas power amplifiers produce large current gain but smaller voltage gain. For optical amplifiers, the distinction seems to depend on the power level of the output beam (high power beam = power amplifier), but it is relatively ill-defined.

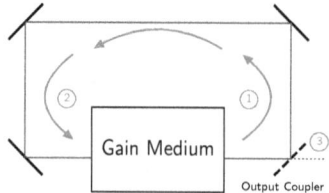

Figure 1.3. Simplified schematic of a laser oscillator. The red lines show laser beams in the cavity (formed by mirrors shown in black), and the blue arrows show the direction of travel around the cavity. Circled numbers show important areas of the cavity, see text for details. The oscillator pump beam is not shown.

The function of the oscillator shown in figure 1.3 can be described as follows (following the circled numbers). (1) Emission (generally stimulated) from the gain medium is directed into the cavity using mirrors. (2) This emission travels around the cavity and is directed back into the gain medium, causing further stimulated emission, and oscillation. (3) The first mirror after the gain medium is an **output coupler**, a partially reflective mirror that reflects some light back into the oscillator (as in stage (1)), but allows some light out to be amplified further in the laser system.

Ensuring that the oscillator is oscillating in the desired way (so it is producing the pulses you want) is a key design consideration. A full mathematical description of how this is achieved is actually quite accessible and can be found in (for example) references [1, 2, 5], but is beyond the scope of this book. Essentially, there will often be multiple competing processes taking place within the oscillator cavity, producing different kinds of laser output. When the light makes one complete circuit around the cavity there will be a certain amount of gain, referred to as the **round-trip gain**; but also a certain amount of loss, referred to as the **cavity loss**. Intuitively, if the round-trip gain is greater than the cavity loss, then the laser will produce output. If, conversely, the cavity loss is greater than the round-trip gain, then there will be no laser output[6]. The point at which the round-trip gain exceeds the cavity loss is referred to as the **lasing threshold**. The construction and alignment of the oscillator has to be such that the desired process has the highest round-trip gain and lowest losses, so that it dominates over other undesired processes. How this is practically achieved in ultrafast oscillators is discussed in chapter 5. The weak laser beam from the oscillator may pass through several stages of subsequent amplification before it is used for the desired application. We will discuss the workings of more specific ultrafast oscillators in due course.

An analogous system which you may be more familiar with arises if you are connecting an MP3 player to a set of speakers. The MP3 player itself cannot drive loudspeakers (it is the oscillator), so the signal has to go through subsequent amplification stages (preamps and power amps) so that it is able to efficiently drive a loudspeaker. This is exactly analogous to how weak pulse from an oscillator is amplified into experimentally usable laser pulses using a laser amplifier.

---

[6] The output will be dominated by spontaneous emission rather than stimulated emission, so will not produce usable laser light.

References

[1] Hooker S and Webb C 2010 *Laser Physics* 1st edn (Oxford: Oxford University Press)
[2] Milonni P W and Eberly J H 2010 *Laser Physics* 1st edn (New York: Wiley)
[3] Weiner A M 2010 *Ultrafast Optics* 1st edn (New York: Wiley)
[4] Paschotta R P 2008 *Field Guide to Lasers* 1st edn (Bellingham, WA: SPIE Press)
[5] Paschotta R P 2008 *Field Guide to Laser Pulse Generation* 1st edn (Bellingham, WA: SPIE Press)

Chapter 2

Laser light and laser beams

Having briefly discussed how lasers work, now we turn our attention to the nature of the light that is produced by a laser. Laser light is produced in a confined *beam*, and we will consider both the nature of the produced light, and the beam it travels in. Initially we consider the nature of the laser light itself, rather than the beam that it travels in.

2.1 Laser light

Looking back at the diagram in figure 1.1, we can consider two possibilities. One is that the pump radiation is kept on continuously—this will lead to 'continuous-wave' (cw) laser output, or a **cw-laser**. In this case the laser (ideally) produces light of a single, well-defined frequency, such that the output laser field looks like a perfect sine wave.

Alternatively, the pump radiation can be fired in short bursts. This leads to an output laser field that consists of a series of pulses, and this is called a **pulsed laser**. Many lasers are capable of operating in either pulsed or cw mode, but all ultrafast lasers are pulsed lasers by definition. We will only be discussing pulsed lasers from now on.

2.1.1 Pulsed lasers and time-bandwidth products

It is now useful to remind ourselves of the general time–energy uncertainty relation, a recasting of the familiar quantum-mechanical position–momentum uncertainty relation:

$$\Delta E \Delta t \geqslant \frac{\hbar}{2} \tag{2.1}$$

This relation tells us that a state that has a very well-defined energy (small ΔE) must exist for a very long time (large Δt), and vice versa. This has profound implications for our pulsed laser, as if the laser output is **pulsed**, then by definition it exists for a finite length of time—so Δt is small. This means that ΔE is relatively large for our

doi:10.1088/978-0-7503-3659-8ch2

laser pulse, and there is uncertainty in the wavelength (colour) of the pulse. This can be simplistically thought of as the emission coming from a broad band of levels (as drawn in figure 1.1), rather than a single level with a well-defined energy. **The principle you learnt at school about lasers being monochromatic is, at best, an oversimplification.**

Equation (2.1) is true in a general sense, but we need to be a bit more precise about how exactly we define ΔE and Δt for our laser pulses, since these depend on the temporal and spectral shape of the pulse. Generally ultrashort laser pulses from Ti:Sa lasers are well described as **Gaussian pulses**, with a Gaussian profile[1] in the time domain. We can then identify the width of our Gaussian pulse in the time domain as $\Delta \tau$; this is defined as the full-width-at-half-maximum (FWHM) of the pulse, as illustrated in figure 2.1 (the blue curve). The quantity $\Delta \tau$ is called the **pulse duration**.

To find the width in the frequency (energy) domain—normally referred to as the **bandwidth**—we take a Fourier transform of our time-domain pulse to find the shape of the pulse in frequency space (or, in a lab, we use a spectrometer to measure the frequency spectrum directly). This is the orange curve in figure 2.1. We then find the FWHM of this frequency domain pulse[2] to determine the spectral width $\Delta \omega$. This bandwidth can also be called the 'spectral width', as it is the width of the frequency

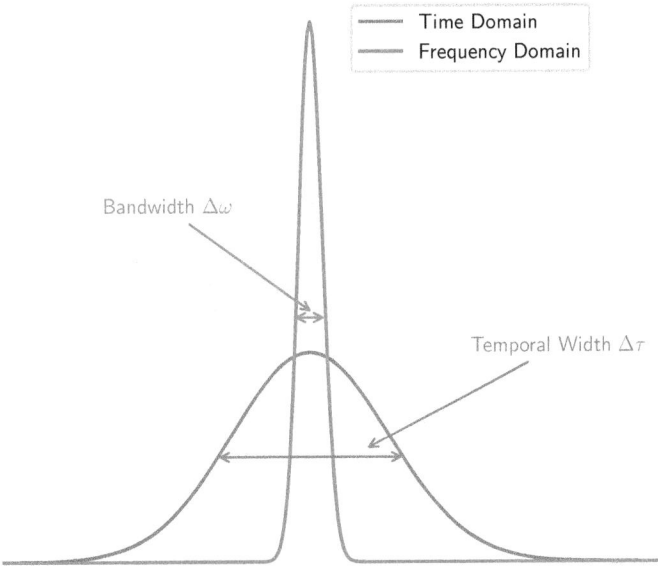

Figure 2.1. A typical Gaussian shaped laser pulse represented in both the time domain (blue) and frequency domain (orange). The frequency domain profile is obtained via Fourier transform of the temporal profile. The definitions of the temporal width and bandwidth are annotated.

---

[1] Different types of laser oscillators produce different pulse shapes. Another common pulse shape is the $sech^2$ pulse shape.

[2] The Fourier transform of a Gaussian is also a Gaussian, so we can find the FWHM in the frequency domain exactly as we did in the time domain.

spectrum that our pulse spans. If there are many colours present in our pulse, then it has a broad bandwidth, and vice versa. An important point to note here is that the spread of frequencies is not, in reality, continuous as figure 2.1 suggests. In a very well-resolved spectrum we would see many individual frequencies in our total spectrum. This arises because generally the light has to form a *standing wave* within the oscillator cavity, so we can **only** have frequency components present where the wavelength is such that an integer number of waves will exactly fit into our cavity length. Mathematically this means that:

$$\text{Wavelength of an allowed component} = \frac{\text{Cavity Length}}{\text{Integer}}$$

So in reality we would see wavelengths at integer fractions of our cavity length, if we had a powerful spectrometer capable of resolving them. Overall, the gain medium produces a wide range of different frequencies as determined by the emission bandwidth. Only a subset of these are allowed by the cavity length, and we will see that it is the **phase relationship** between these frequencies that produce the ultrashort pulse. The interplay between these effects is illustrated in figure 2.2.

Turning back to a discussion of bandwidth, we can now use our new definitions $\Delta\omega$ and $\Delta\tau$ to define a new kind of uncertainty relation in terms of the product $\Delta\omega\Delta\tau$. This is referred to as the **time-bandwidth product** for obvious reasons. For a Gaussian pulse, it is found that:

$$\Delta\omega\Delta\tau \geqslant 0.441 \tag{2.2}$$

This means that for a Gaussian pulse with a given bandwidth $\Delta\omega$, the shortest possible pulse is found when this inequality is an equality, and then $\Delta\tau = 0.441/\Delta\omega$.

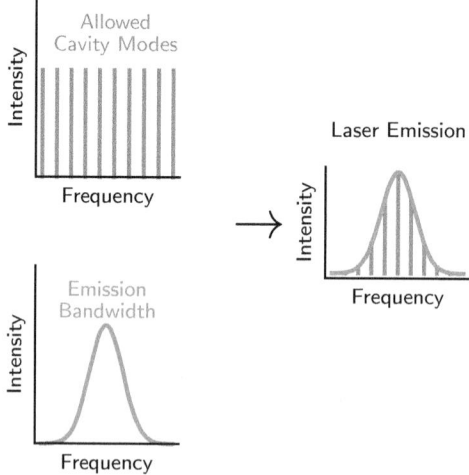

Figure 2.2. Illustration of how the interplay between the modes allowed by the cavity (top left, blue) and the emission bandwidth of the gain medium (bottom left, orange), give rise to the overall laser emission spectrum (right). Note that in most real cases the individual cavity frequencies shown in blue would not be individually resolvable.

An important point to note here is that equation (2.2) originates ultimately from the properties of the Fourier transform, even though the physical origin of emission broadening comes from the Uncertainty Principle. Fourier theory predates quantum mechanics, and arguably Fourier was describing the same situation as Heisenberg, but was unable to provide a physical explanation. Heisenberg's Uncertainty Principle provides a physical basis for these properties of the Fourier transform.

To take a rough illustrative example, if we take a typical bandwidth of a Ti:Sa laser of 60 nm (or 28 THz at a central wavelength of 800 nm), then the shortest possible pulse this bandwidth could support is theoretically around 15 fs![3] When the temporal width of the pulse is as short as the bandwidth will allow, we say that the pulse is **transform-limited**, **Fourier-limited**, or that it is **at the transform limit**.

2.1.2 The transform limit

Our time-bandwidth product shows that a laser pulse with a broader bandwidth can have a shorter temporal width than one with a narrower bandwidth. We normally talk about the shortest pulse possible with a given bandwidth as being the shortest pulse that the bandwidth can **support**, and this shortest pulse occurs when the pulse is **transform-limited**. A clear question then, is *'are pulses with a broad bandwidth necessarily short?'* The answer is a resounding **no**, and the part of the remainder of this chapter and the next will explore why.

To understand the concept of transform-limited pulses in more detail, it is useful to think about the pulse in the frequency domain. The pulse has a bandwidth $\Delta\omega$, which is spread around a central frequency ω_0. Within this bandwidth there are many different **spectral components**. For example, in the case of an 800 nm pulse (374 THz) with a bandwidth of 60 nm (28 THz), we would expect to see components between 770 nm (390 THz) and 830 nm (361 THz) if we looked at the spectrum of our pulse. Each of these components can be considered as a sine wave with electric field $E(t)$:

$$E(t) = E_0 \sin(\omega t + \phi(\omega)),$$

where the frequencies ω of the waves are distributed around ω_0 in accordance with the bandwidth (larger bandwidth → larger spread of frequencies). Here $\phi(\omega)$ is the **phase** of the wave with frequency ω, and E_0 is the amplitude of the electric field. The quantity $\phi(\omega)$ is often called the **spectral phase**, as it is the phase of the spectral component with frequency ω. The phase of a wave tells you what part of the cycle of the wave you are in, and is discussed further in appendix A if this is unfamiliar. It is the *phase relationship* between different frequency components (that is, the difference in their phases $\Delta\phi$) that is critical to understanding ultrafast pulse generation. A simple example that illustrates why phase is important is shown in figure 2.3—in which two waves are interfered. In the top frame, both waves are perfectly in phase and have a phase difference $\Delta\phi$ of zero, leading to constructive interference when the

---

[3] This will seem very short to anyone with experience, and is because not all of this bandwidth can be efficiently amplified or compressed.

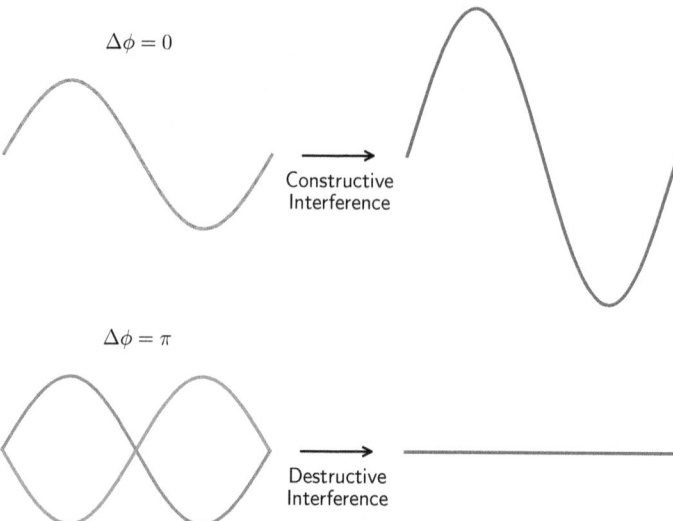

$\Delta\phi = 0$

Constructive
Interference

$\Delta\phi = \pi$

Destructive
Interference

Figure 2.3. Effects of different phase relationships on the interference of two waves (orange and blue), with zero phase difference (top) and with a phase difference π radians (bottom). The resultant wave is shown in red in both cases.

amplitudes are summed. In the lower frame, both waves are out of phase by π radians, which leads to destructive interference.

In a typical Ti:Sa laser oscillator, we may have around 250 000 colours (waves of different frequencies) all resonating around the cavity. If each wave has a well-defined phase relationship to all the other waves, then the waves are said to be **phase-locked**, and the cavity is said to be **mode-locked**. This results in constructive interference and creation of ultrashort laser pulses. Conversely, if all the phases are random, and there is no well-defined phase relationship between one wave and the others, then on average there is destructive interference—and no pulse is seen. This is illustrated in figure 2.4.

Creating the mode-locked cavity is key to generating ultrashort pulses, and we will discuss how to do this in chapter 5, but the broader point to take from this is that the more modes we have that are phase-locked, then the shorter our output pulse can be. When all of the frequency components have a well-defined phase relationship[4], then they will temporally coincide at a certain point; and this situation is the transform limit. **If all of the colours in the pulse arrive at the same time, the pulse is transform-limited.** This concept will be explored more mathematically in the next chapter, but it is illustrative to consider the effect of adding a greater number of frequencies together on the output pulse. This is shown in figure 2.5, and it is clear that adding more frequencies together in phase creates a shorter output pulse. In fact, the only reason that we see any pulsed output at all is because we add together many different frequencies in phase.

---

[4] This is equivalent to saying that the $\phi(\omega) = c_n\omega$, where c_n is a constant specific to each frequency. Then $\frac{d\phi}{d\omega} = c_n$, and the phase difference between different components does not depend on the frequency.

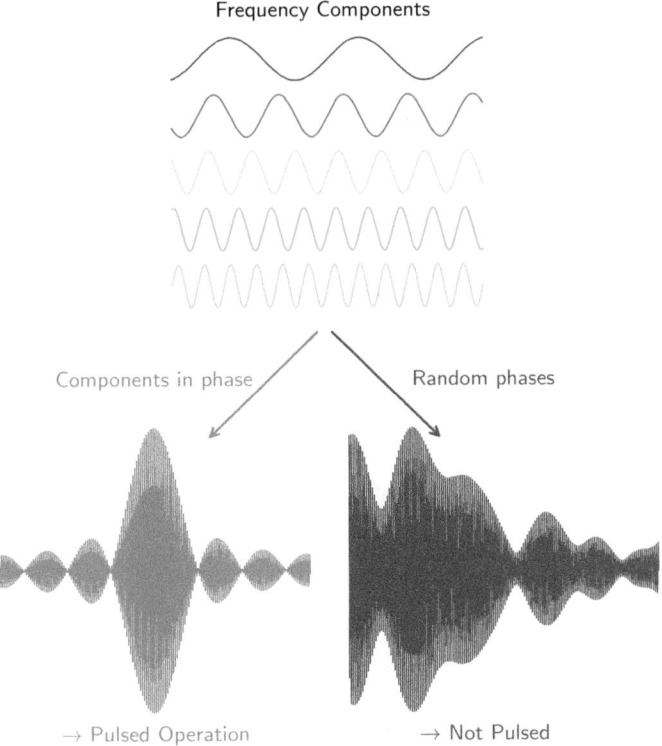

Figure 2.4. Adding waves of different frequencies with a well-defined phase relationship leads to generation of a pulse, whereas adding waves with random phases does not. The output waveforms are calculated by superimposing ten waves from within a typical 800 nm Ti:Sa bandwidth, whereas the initial plane waves are chosen just for illustrative reasons.

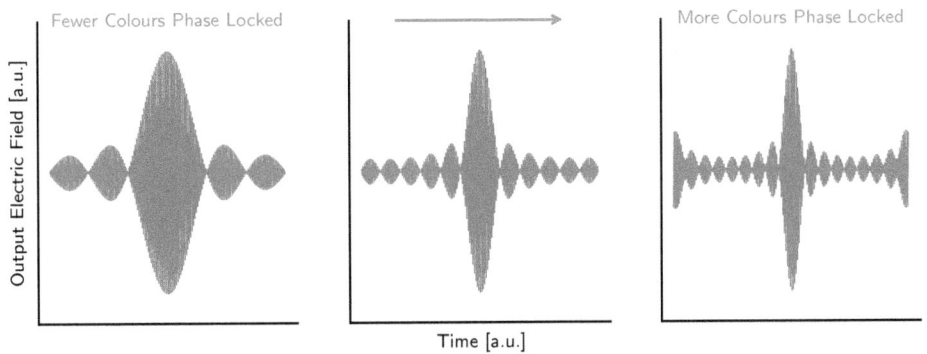

Figure 2.5. Adding together fewer frequencies in phase (left) gives a relatively broad output pulse. Adding more frequencies in phase (middle and right) produces a narrower output pulse. This lies at the heart of ultrashort pulse generation.

2.1.3 Repetition rates and pulse trains

In the preceding sections we have discussed the properties of ultrafast laser light, and seen that the light exists as short pulses of EM radiation. The laser system will produce a certain number of these pulses every second, and this number is known as the **repetition rate**, or **rep rate**, of the laser system (measured in Hz). Thus, from the laser output we get a train of the output pulses known as a **pulse train**.

Repetition rates can vary from <1 Hz for very large and powerful lasers, up to many hundreds of MHz for small oscillators. Lasers producing very high pulse energies tend to have lower repetition rates and vice versa. Most tabletop femtosecond systems have a repetition rate between 1 kHz and 1 MHz. The **average power** of a laser (in Watts) is defined using the rep rate as follows:

$$\text{Average Power (W)} = \text{Pulse Energy (J)} \times \text{Rep Rate (Hz)}$$

The *average power* of the system is distinct from the *peak power* of a single pulse. We will discuss this further in the context of beam characterisation in section 6.3.

2.1.4 Polarisation

Having discussed the temporal and spectral characteristics of our ultrafast laser output, there is one more characteristic of the light to discuss before discussing the characteristics of the laser *beam* itself. Laser light is simply an oscillating electric field travelling in a beam with a well-defined direction (see appendix A for a discussion of travelling EM waves). The direction of travel of the beam is referred to as the **propagation direction**, and the direction of oscillation of the electric field must be orthogonal to the propagation direction. This arises from **Maxwell's equations**, which are beyond the scope of this work, but an excellent accessible introduction to Maxwell's equations and vector calculus can be found in references [1, 2]. The direction that the electric field oscillates in is known as the *polarisation* of the laser light. Polarisation is a vector that points in the direction of the field oscillation. The plane that is orthogonal to the direction of travel contains the polarisation vector(s) and is therefore known as the **polarisation plane**. This is illustrated in figure 2.6 for a linearly polarised wave.

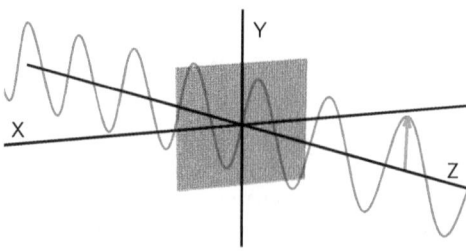

Figure 2.6. Coordinate frame with an illustrative example of polarisation of a light wave (blue). The wave is propagating along the Z-axis, and the polarisation plane is the XY plane (shaded brown). In this instance, the light is linearly polarised along the Y direction, and the polarisation direction is shown with the orange arrow.

Light can be linearly, elliptically, or circularly polarised within the polarisation plane, depending on the exact way in which the field oscillates. More detail on how each of these polarisation states form can be found in any introductory text on optics [3, 4], and is not included here. Polarisation in ultrafast optics is essentially no different to polarisation in non-ultrafast optics. There are, however, some aspects of polarisation nomenclature which we will use going forward, which are documented below for clarity.

Defining the polarisation direction

Linearly polarised light is most commonly used, and there are a couple of different systems that are used to refer to the polarisation state of the light. You will hear people talk about 'vertical', 'horizontal', 's-polarised', or 'p-polarised' light.

The vertical/horizontal nomenclature seems at first most intuitive. The direction of the field oscillation is specified with respect to its orientation to the surface of the Earth (and laser tables are normally parallel to the surface of the Earth!). Then, if the light is polarised such that the direction of oscillation is parallel to the plane of the Earth/table, it is said to be **horizontally polarised**. If it is polarised such that the direction of oscillation is perpendicular to the plane of the Earth/table, it is said to be **vertically polarised**. These designations are normally very convenient, as when standing in a lab, 'vertical' means 'vertical' with respect to your laser table.

However, this designation becomes more challenging if a beam is not propagating parallel to a table (for example, if you are sending a beam upwards in a periscope). Then the entire polarisation plane is parallel to the table, and so all light would be 'horizontally polarised' by this definition. To remedy this issue, the **s and p polarisation** designations are used.

The s and p polarisation designation defines the polarisation of an incoming wave with respect to a plane known as the *plane of incidence*, which is the plane spanned by the surface normal of an optical element (such as a mirror) and the propagation direction of the incoming wave. More simply, it can be thought of as the plane that the beam travels in both before and after reflection or refraction. If the polarisation direction is parallel to this plane, the light is said to be **p-polarised** ('p' for parallel). Conversely, if the polarisation direction is perpendicular to this plane, the light is said to be **s-polarised** ('s' from the German 'senkrecht', meaning perpendicular). You can see that when a beam is propagating parallel to a laser table, then s-polarised light is vertically polarised and p-polarised light is horizontally polarised. The advantage of the s/p designations is that it makes defining the polarisation when the beam is *not* propagating parallel to the table much less ambiguous. This is illustrated in figure 2.7.

Figure 2.7(a) shows the specific case where the beam is travelling parallel to the table. In this case, the plane of incidence is parallel to the laser table, and p-polarised light is always 'horizontally polarised', and s-polarised light is always 'vertically polarised'. Both systems of naming work equally well in this case. However, figure 2.7(b) shows the specific case where a beam is propagating perpendicular to the table. At this point where the beam is travelling perpendicular to the table, the plane of incidence is perpendicular to the table, and both s and p polarisation states

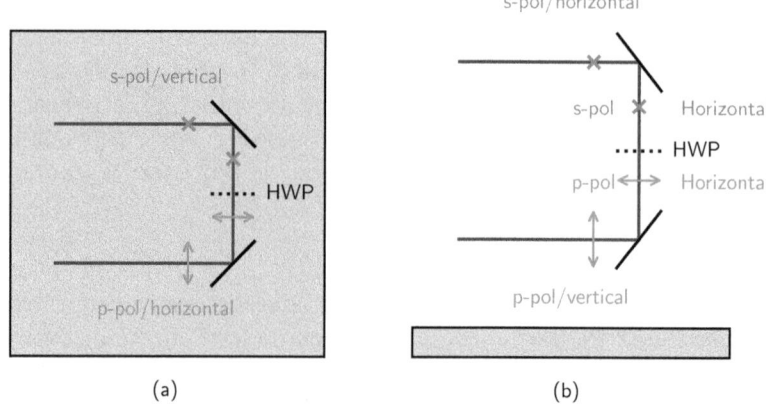

(a)   (b)

Figure 2.7. Illustration of different ways to describe polarisation of a given laser beam (red), which is reflected off two mirrors and has its polarisation rotated by a waveplate between the mirrors. The polarisation state is shown as an orange arrow or blue cross, the cross denoting polarisation into the plane of the page. (a) Beam parallel to the table. (b) Beam perpendicular to the table.

would be considered 'horizontal' to the table. The horizontal/vertical designation becomes ambiguous in this case. Using the s/p designation removes this ambiguity, so you will normally find that when (for example) optics manufacturers quote polarisation dependent metrics for their optics, they will use the s/p designation.

2.2 Gaussian beams

We now turn to discuss the nature of the *beam* that the laser light travels in, rather than the nature of the light itself. Almost all laser beams you will come across are well described as **Gaussian beams**. A Gaussian beam is a beam where the amplitude of the beam in the *transverse plane* is described by a Gaussian function. The **transverse plane** here refers to the plane perpendicular to the propagation direction of the beam, i.e. along the 'face' of the beam[5].

This discussion is, like that of polarisation, not specific to ultrafast lasers. However, it is of critical importance for an experimentalist to have some knowledge of Gaussian beams, especially when it comes to characterising laser beams. We will focus on knowledge that helps us use and manipulate Gaussian beams; many standard texts on laser physics will give more information as to *why* and *how* Gaussian beams are produced by laser oscillators [5, 6].

2.2.1 Ideal beam parameters

In an ideal world[6], when we look at the output of our laser beam on a card, we will see a circular spot that trails off towards the edges. Ideally, and in the simplest case,

---

[5] The same plane as the polarisation plane.
[6] Deviations from this ideality will be briefly discussed in due course.

the variation in **intensity** I of this spot as a function of radius r from the centre of the spot is given by equation (2.3).

$$I(r) = I_0 \exp\left(-\frac{2r^2}{x^2}\right),\qquad(2.3)$$

where I_0 is the **peak intensity**, and x is known as the **beam radius**[7]. **Intensity** is a physical quantity that simply refers to the **power per unit area** transferred by something, in this case a laser beam. The beam radius is defined as the distance away from the centre of the spot (distance from the axis of propagation of the beam) where the intensity falls to $1/e^2$ of I_0. This is illustrated in figure 2.8.

As figure 2.8 shows, the vast majority of the beam intensity is enclosed within the $1/e^2$ radius; about 86% of the power in the beam will pass through a circle with the $1/e^2$ beam radius.

However, what we have measured by looking at the beam on our card in this way is only the beam radius *at the point where we measured it*. Laser beams are usually not perfectly **collimated**, that is, they normally **diverge** or **converge** as they propagate. The extent to which the beam diverges or converges is called the **divergence** of the beam, and this means the beam radius will *vary*, depending on

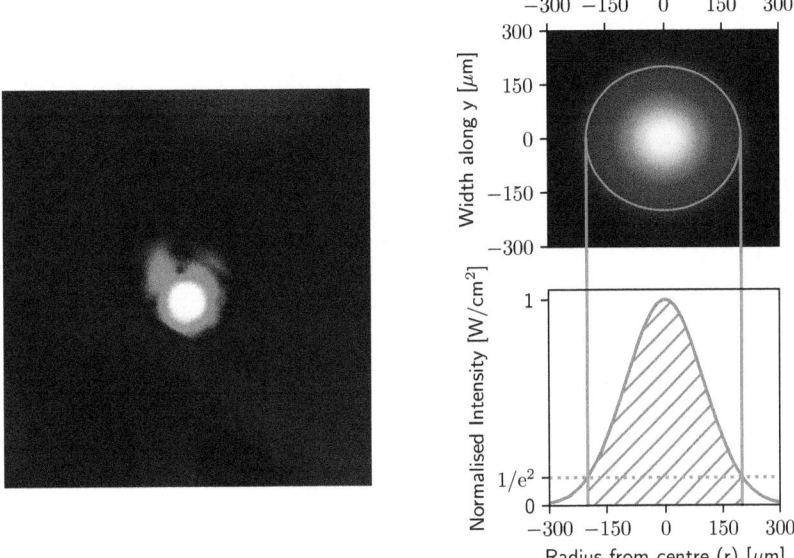

Figure 2.8. (a) A photograph of a typical laser output, taken in the author's lab. (b) Simulation of a circular 2D Gaussian laser spot with a beam radius of 200 μm (top), and the corresponding radial projection (bottom). The orange circle and lines to the projection denote the area enclosed by the $1/e^2$ beam radius. This is also shown by the blue hatched area on the projection.

---

[7] This is more commonly referred to as ω, but to avoid confusion with frequency, we use x here.

where in the beam path we measure it. The beam divergence θ is defined in a fairly intuitive way:

$$\theta = \arctan\left(\frac{x_2 - x_1}{L_{21}}\right) \tag{2.4}$$

Strictly, this is the *radial* beam divergence (the 'divergence half-angle'), and sometimes you will see it defined with a capital $\Theta = 2\theta$ (the 'divergence full-angle'). Qualitatively, we can explain equation (2.4) simply by saying that we measure the beam radius ω at two positions 1 and 2, and the distance between these positions L_{21}. Straightforward trigonometry then leads to the definition of θ given. These concepts are illustrated in figure 2.9 in green. The lowest possible divergence for a Gaussian beam (i.e. the divergence of an ideal Gaussian beam) is given by:

$$\theta = \frac{\lambda}{\pi x} \tag{2.5}$$

So longer wavelength beams diverge more than shorter wavelength beams, and larger beams diverge less than smaller beams, but all beams will diverge to some extent.

Normally a beam taken straight from the output of a good-quality laser (i.e. not a laser pointer) will have very low divergence, and so the beam radius doesn't vary appreciably except over very large distances. As an example, an ideal Gaussian beam from an 800 nm laser with an initial beam diameter of 10 mm would have to propagate for over 50 m before the beam diameter gets bigger than 11 mm. However, if we **focus** our beam, then the divergence changes hugely, and the size of the beam radius will decrease drastically as we measure it nearer and nearer to the focus. This is shown in figure 2.9 as the orange lines, which are the outline of a Gaussian beam which is focussed at the centre of the plot (and diverges away from the centre). In the case that the beam diverges or converges, it's not clear how we should define our beam according to equation (2.3). To account for this variation of

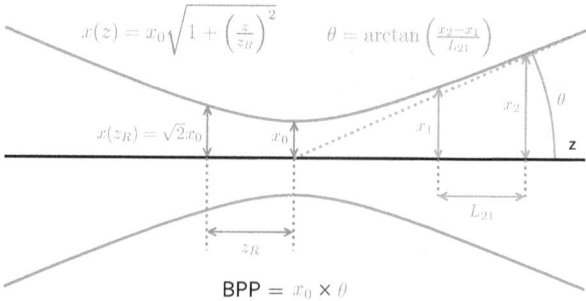

Figure 2.9. Illustration summarising the various parameters used to describe a generic Gaussian beam. The beam profile (orange) is shown side-on. Parameters relating to the beam waist x and Rayleigh length z_R are shown in blue. Parameters relating to the divergence angle θ are shown in green.

the beam radius as the beam propagates, we have to refine equation (2.3) such that the beam radius $x \rightarrow x(z)$, where z is the distance away from the point in the beam path where the radius is smallest.

$$I(r, z) = I_0 \exp\left(-\frac{2r^2}{x(z)^2}\right) \tag{2.6}$$

When the beam radius is at its smallest, it is known as the **beam waist**, and is given the symbol x_0. It can also be shown that the variation in beam radius $x(z)$ can be expressed as:

$$x(z) = x_0\sqrt{1 + \left(\frac{z}{z_R}\right)^2}, \tag{2.7}$$

where z_R is known as the **Rayleigh range** or **Rayleigh length**. These concepts are illustrated in figure 2.9 in blue. At a distance of z_R from the position of the beam waist, the beam radius has increased by a factor of $\sqrt{2}$; the Rayleigh length therefore gives some measure of the range over which the beam remains small—i.e. if we have focussed the beam, how long is the region where the beam is still effectively 'focussed'?

The beam waist x_0, divergence θ, and Rayleigh length z_R are three of the most useful parameters when characterising Gaussian beams, and are readily measured in the lab. In fact, the quality of a Gaussian beam is defined using the **beam parameter product (BPP)**. This is defined as the product of the divergence θ and beam waist x_0:

$$\text{BPP} = \text{Beam Waist } (x_0) \times \text{Divergence half–angle } (\theta)$$

The units of θ and x are normally given in mrad and mm respectively; so the units of the BPP are mm mrad. Comparison with equation (2.5) readily reveals that the BPP of an ideal Gaussian beam is simply λ/π. A beam with this BPP will have the smallest possible beam waist for a given divergence, and vice versa.

The more practical ramifications of beam waists, divergences, Rayleigh lengths, and BPPs, such as how you can measure them, and how you can use them, are discussed in a later section. However, we made reference to the preceding discussions only being valid in 'ideal' conditions—what these conditions are is the subject of the next section.

2.2.2 Deviations from ideality

Beam quality factors—M^2
In our discussion of laser oscillators in subsection 1.3.2, we mentioned that oscillators can produce different kinds of laser output, but that the construction of the oscillator was such that it would suppress undesired output and enhance the desired output. One way in which oscillators can produce a mixture of different desired/undesired outputs is in the production of different **transverse modes** of the laser output. A transverse mode is essentially 'what the front of the laser beam looks like', that is, the shape of the 'face' of the beam. The idea of a *longitudinal mode* in a

standing wave may be familiar from school physics, such as when a guitar string oscillates. The difference between longitudinal and transverse modes is shown in figure 2.10.

Figure 2.10(a) shows longitudinal modes of a standing wave in a cavity. Here the direction of travel of the wave is in the plane of the page (we are looking at the wave 'side-on'). This situation can be seen/experienced regularly when guitar strings vibrate, or when a flute is played. Figure 2.10(b) shows transverse modes of a Gaussian beam. Here the direction of travel of the wave is out of the plane of the paper (we are looking at the wave 'head-on'). The modes are labelled with the 'TEM $_{m,n}$' nomenclature; TEM stands for 'Transverse ElectroMagnetic' mode, and the subscripted numbers refer to the degree of the *Hermite polynomial* in the mathematical expression of the mode. The precise mathematical form of the modes need not detain us here, and we will focus on the important point, which is that **ideally, our laser will output light only in the lowest, TEM$_{00}$, transverse mode.**

If our laser is not outputting light in only this lowest mode, and there are contributions from higher order modes present[8], then our beam is no longer an ideal Gaussian beam. This means that our BPP will be greater than the ideal value of λ/π. Intuitively, we could get a qualitative feel for *how* non-ideal our beam is by taking the ratio of the BPP for our beam to the BPP for an ideal Gaussian beam of the same wavelength. This ratio has a name, the **beam quality factor**, or **M^2 ('M-squared')** factor.

$$\text{Beam Quality Factor } (M^2) = \frac{\text{BPP of real beam}}{\text{BPP of ideal Gaussian beam}}$$

As we are only considering Gaussian beams, we know that the lowest possible BPP is for that of an ideal TEM$_{00}$ Gaussian beam. This means that $M^2 \geqslant 1$. If $M^2 = 1$, then the beam is ideal. You may also hear people talk about a beam being

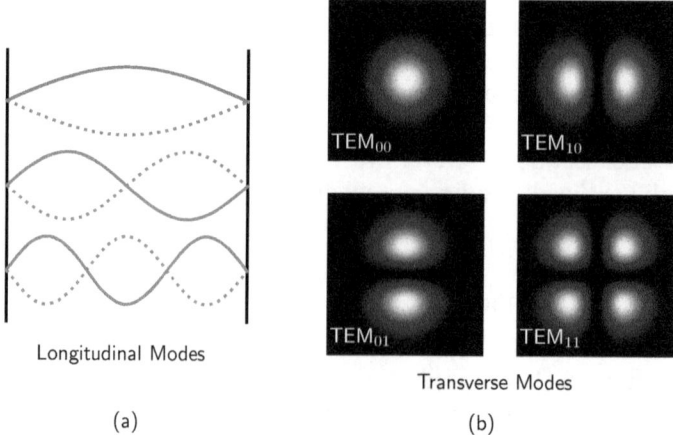

Longitudinal Modes

Transverse Modes

(a)   (b)

Figure 2.10. (a) Lowest three *longitudinal modes* of a standing wave in a cavity. (b) Lowest four *transverse modes* of a Gaussian beam.

---

[8] This can occur due to poor oscillator alignment, for example.

diffraction-limited[9], this is equivalent to saying that $M^2 = 1$. In practice, provided that $M^2 < 1.2$, the beam can be considered as behaving reasonably ideally.

We will discuss the practical implications of this, and how you can measure the M^2 value, in chapter 6. For now, just being aware of the terminology will be useful as you start to work in the lab and think about your beams.

Elliptical beams

Another area where our assumptions of ideal behaviour in subsection 2.2.1 break down is if our beam spot is **elliptical**, that is, is not circular. In this instance, we cannot define the beam radius as we had previously—as the radius will vary depending on where on the ellipse we look at the beam! In this instance, the normal thing to do is to quote the radius at the widest and narrowest parts of the ellipse[10]. Then, rather than quoting the beam radius/waist as a single number, it would be quoted as two numbers, for example '25 μm by 35 μm'.

Sometimes elliptical beams are formed deliberately using cylindrical lenses, as they can be advantageous in some spectroscopy experiments. Often they are undesirable, but form anyway due to imperfections in optics in a beam path, or imperfections in the laser output. Dealing with elliptical beams can be difficult, but most often they form when an optic is over-tightened in a mount (and so deforms slightly producing a lensing effect), or when a focussing optic isn't placed exactly square in the beam path. Checking that all your telescopes are properly aligned, and that no thin optics are over-tightened can remedy the situation. Sometimes a beam is only very slightly elliptical, such that just taking the mean of the two beam radii gives a reasonable number for an 'effective' circular beam radius that can be used in subsequent calculations.

We have now summarised all of the useful properties of laser light and laser beams that we will need going forward. In the next chapter, we will return to our discussion of transform-limited pulses, and understand when our ultrashort pulses are *actually* ultrashort, and when they are not.

References

[1] Fleisch D 2008 *A Student's Guide to Maxwell's Equations* 1st edn (Cambridge: Cambridge University Press)

[2] Schey H M 2005 *Div, Grad, Curl, and All that: An Informal Text on Vector Calculus* 4th edn (New York: W W Norton)

[3] Pedrotti F L, Pedrotti L M and Pedrotti L S 2017 *Introduction to Optics* 3rd edn (Cambridge: Cambridge University Press)

[4] Goldstein D H 2017 *Polarized Light* 3rd edn (Boca Raton, FL: CRC Press)

[5] Milonni P W and Eberly J H 2010 *Laser Physics* 1st edn (New York: Wiley)

[6] Hooker S and Webb C 2010 *Laser Physics* 1st edn (Oxford: Oxford University Press)

---

[9] This terminology arises because at the diffraction limit, the beam waist cannot be lowered any further. Attempting to reduce the beam waist below the diffraction limit will lead to broadening of the beam waist due to diffraction.

[10] The semi-major and semi-minor axes of the ellipse.

IOP Publishing

Ultrafast Lasers and Optics for Experimentalists

James David Pickering

Chapter 3

Dispersion

In the previous chapter we described some of the important properties of laser light, but didn't really fully answer the question we set out to answer, which was 'are pulses with a broad bandwidth necessarily short?' We stated that to have a transform-limited pulse (the shortest possible pulse for a given bandwidth), there had to be a fixed phase relationship between all the frequency components in the pulse, such that at some point all of the different components (different colours) temporally coincide to produce a short pulse.

Dispersion is the phenomenon which stops the different frequency components arriving at the same time, and leads to broadening of the pulse. Dispersion is a phenomenon which is critical in ultrafast optics. If you are only familiar with non-ultrafast optics then you may have never really thought about it, and it is the single biggest difference that we have to be mindful of when moving from non-ultrafast optics to ultrafast optics. Dispersion, unless properly taken care of, can cause ultrafast pulses to no longer resemble anything even close to ultrafast pulses—so it is a very important concept to understand.

An initial disclaimer is that when we refer to 'dispersion' we are strictly referring to **chromatic dispersion** or **temporal dispersion**—the spreading out of different colours *in time*. There are other kinds of dispersion too, such as **spatial dispersion**, which will we mention later. Whenever we refer to an unqualified 'dispersion', we are referring to chromatic dispersion.

3.1 Origins of dispersion

The **refractive index**, n, of a material is defined as:

$$n = \frac{c_0}{c},\tag{3.1}$$

where c_0 is the speed of light in vacuum, and c is the speed of light in the material. In most materials, $c < c_0$ and therefore $n > 1$. Moreover, the refractive index of a

doi:10.1088/978-0-7503-3659-8ch3

material is frequency dependent, i.e. $n = n(\omega)$. **This is the physical origin of dispersion.** If different colours of light experience different refractive indices, then (for example) the higher frequency 'bluer' colours could travel more slowly through the material than the lower frequency 'redder' colours. This situation in which higher frequency light travels more slowly is called **normal dispersion**. The opposite situation, when higher frequency light travels faster, is called **anomalous dispersion**. These two cases are summarised in table 3.1. In both cases, dispersion will cause a transform-limited short pulse (which necessarily contains many colours) to broaden in time as it passes through a medium.

The wavelength dependence of the refractive index for a material can be calculated easily using the *Sellmeier equation* for that material, but for practical use it is much more convenient to use the excellent website https://refractiveindex. info, which contains wavelength dependent refractive index data for a wide range of common optical materials. Figure 3.1 shows the refractive indices of some common optical materials, plotted against both frequency and wavelength. Most materials exhibit normal dispersion, but some show anomalous dispersion in narrow regions[1].

Rearranging equation (3.1), we find that in general the speed at which a given frequency component travels through a medium is given by:

$$c(\omega) = \frac{c_0}{n(\omega)} \tag{3.2}$$

The speed of a particular frequency component in a pulse is known as the **phase velocity** of the component. This is distinct from the **group velocity**, which is the velocity of the entire pulse, as the different frequencies that make up the pulse travel at different speeds through a given medium, according to equation (3.2).

3.2 Dispersion, ultrafast pulses, and chirp

We have seen that the refractive index is frequency (or wavelength) dependent, and that this can cause dispersion as different colours of light travel at different speeds through a given medium. For non-ultrafast laser pulses, this is not usually problematic—as the bandwidth is sufficiently narrow that the effects of dispersion are essentially negligible. When working with ultrashort pulses though, dispersion becomes a critical consideration. This is because an ultrashort pulse **inherently has**

Table 3.1. Summary of different types of dispersion.

Dispersion	Refractive index	Blue light	Red light
Normal	Increases with ω	Travels slower	Travels faster
Anomalous	Decreases with ω	Travels faster	Travels slower

---

[1] Note that some sources will talk about 'anomalous dispersion' to mean 'negative GDD', so it is important to be aware of the context. Many materials have large regions where they exhibit negative GDD, but this does not necessarily mean that the refractive index of the material decreases with frequency.

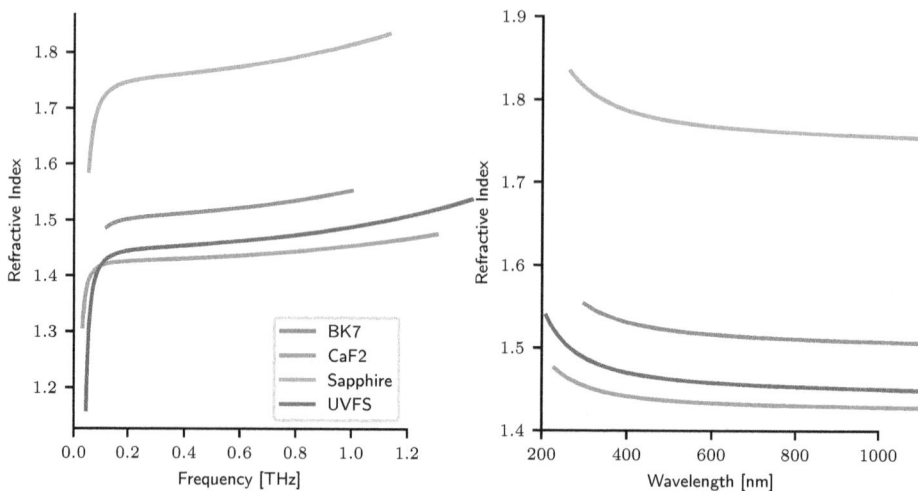

Figure 3.1. Refractive indices of some common optical materials, as a function of frequency (left) and wavelength (right). The plot against wavelength is truncated to cover only the visible and near-IR regions, to illustrate how refractive index changes in this commonly used region. Materials chosen are BK7 (blue); Calcium Fluoride (orange); UV-fused silica (red); and Sapphire (green). Note that not all materials have data available for all wavelengths of light, hence some curves are longer/shorter than others.

a broad bandwidth. This is an important subtlety to understand. It is not directly the short temporal width that causes ultrashort pulses to be affected more by dispersion. Rather, it is that *broadband* pulses are affected more by dispersion, and ultrashort pulses are necessarily broadband.

To qualitatively understand the effect that dispersion can have on an ultrafast pulse, imagine we have a hypothetical transform-limited pulse with a very broad bandwidth, containing frequency components from dark red to deep blue. The pulse is transform limited, so all of the different frequency components arrive at the same time. If, however, we pass this pulse through a piece of glass with normal dispersion, then both the blue and red frequency components slow down, but the red frequency components slow down less than the blue. If we looked at the pulse after it has passed through the glass, then we would see more red components at the front of the pulse (as they travelled faster through the glass), and more blue components at the back of the pulse (as they travelled more slowly through the glass). Therefore, our colours no longer arrive at the same time, and our pulse is **no longer transform limited!** As the different frequency components have been **dispersed** in time, then the pulse **must** have broadened. This effect is more dramatic for pulses with wider bandwidths, and for materials with a higher dispersion.

Following the passage through a medium with positive dispersion, the red frequency components arrive before the blue frequency components—this is the case plotted in figure 3.2[2]. At this point, we say that the pulse is **chirped**. The reason

---

[2] It is admittedly impossible to tell this when there is no time axis present, but time runs from left to right as normal—so the higher frequencies arrive later.

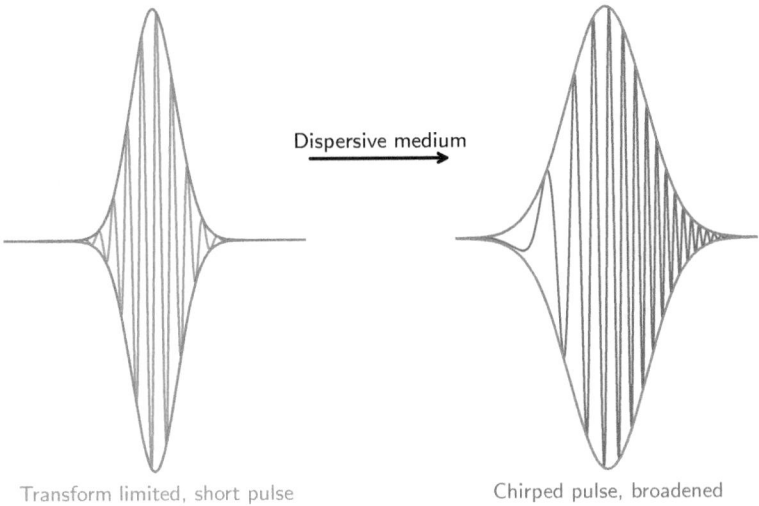

Dispersive medium

Transform limited, short pulse   Chirped pulse, broadened

Figure 3.2. Schematic of dispersion of a TL pulse (orange) passing through a dispersive medium, so it becomes chirped and broadened (red). The extent of the chirp has been exaggerated for illustrative reasons.

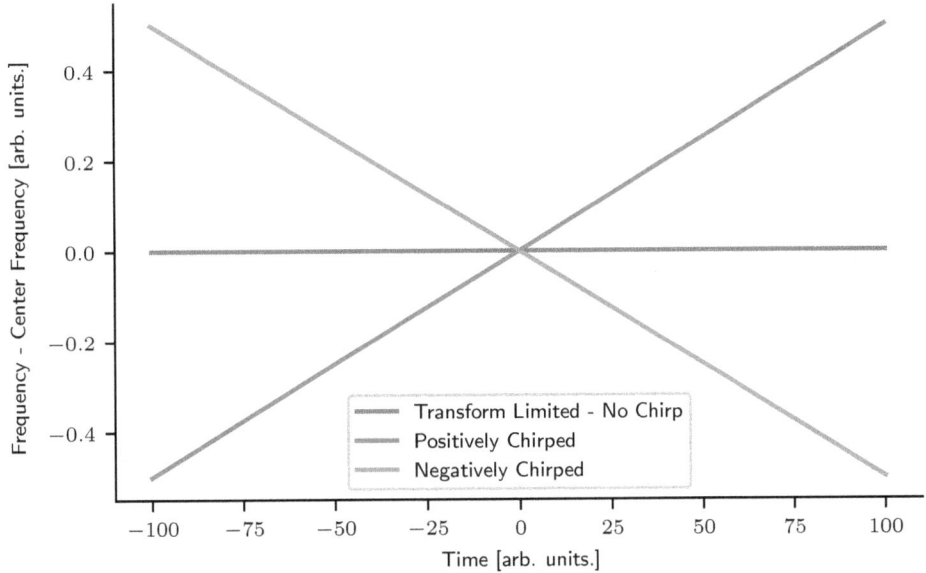

Figure 3.3. Spectrograms for an arbitrary unchirped pulse (blue), positively chirped pulse (orange), and negatively chirped pulse (green). Numerical values are arbitrary and chosen only for illustration.

for this name becomes clear if we look at a spectrogram (plot of frequency against time) for the pulse, as in figure 3.3.

Clearly, for the two chirped pulses shown in figure 3.3 (orange and green lines), the frequency of the pulse changes during the course of the pulse. For a **positively chirped** pulse (orange line), the frequency **increases** during the course of the pulse, because the lower frequencies arrive earlier. For a **negatively chirped** pulse (green line), the frequency **decreases** during the course of the pulse, because the higher frequencies arrive earlier. The name 'chirp' is given, as if you were to downconvert the optical frequencies into audio frequencies, you would hear a rising or falling frequency—rather like a bird chirping. Finally, for the **unchirped** pulse (blue line), the frequency does not change during the course of the pulse—all of the frequencies arrive at the same time.

The next question is to consider how much this added chirp will broaden my pulses? This requires that we calculate how *much* dispersion is added by a particular optical element. This, in turn, requires that we think more mathematically about what happens when light propagates through a dispersive medium.

3.2.1 Envelopes and carriers

Before continuing the discussion, it is useful to understand the concept of the **pulse envelope** and the **carrier wave**. Figure 3.4 shows a sketch of an ultrashort pulse with the pulse envelope (blue) and carrier wave (orange) highlighted for a hypothetical transform-limited pulse.

It is fairly intuitive as to why these are called the carrier and the envelope. The envelope is (in our case) the Gaussian distribution that dictates the temporal shape of the pulse—shown in blue on the figure. The carrier wave is the superposition of all the frequency components present in the pulse. The product of the carrier wave and envelope represents the electric field of the pulse. Generally speaking, the carrier oscillates extremely fast, so much so that the envelope hardly changes from the point of view of the oscillating carrier—so treating them separately in any mathematical analysis is generally a reasonable assumption. This is often called the **'slowly-varying envelope approximation'** (SVEA).

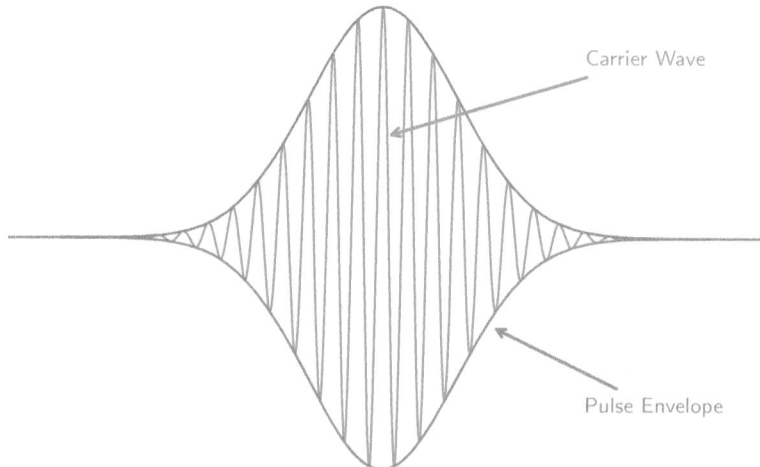

Carrier Wave

Pulse Envelope

Figure 3.4. A hypothetical transform-limited pulse shown with the carrier wave (orange) and pulse envelope (blue).

Note that in the particular pulse shown in figure 3.4, the peak of the carrier wave coincides with the peak of the envelope. When this is the case, we say that there is no **Carrier-Envelope Offset**. In the literature you may encounter this quantity referred to as the 'Carrier-Envelope Phase' (CEP). This is something of importance to the study of **few-cycle pulses**, in which the carrier wave only completes two or three optical cycles within the window provided by the pulse envelope (in this case the SVEA breaks down). It takes a bit of work to make few-cycle pulses, they will not generally be produced straight out of the output of your laser system, and are not considered further here.

3.3 Propagation through a dispersive medium

We will now think more mathematically about the propagation of light through a dispersive medium, to put our prior (and later) discussion on a more formal footing. To avoid dense mathematics getting in the way of an intuitive understanding, this is kept rather brief. More rigorous treatments of this topic can be found in references [1–3].

As our initial discussion of modelocking suggested, a femtosecond laser pulse can be described as a superposition of different frequency components, which all interfere in a way that produces pulsed output (as shown in figure 2.4). We can therefore write the electric field of our pulse in the time domain $E(t)$ as a superposition of different frequencies:

$$E(t) = \frac{1}{\sqrt{2\pi}} \int_{-\infty}^{\infty} \mathcal{E}(\omega)e^{i\omega t}\mathrm{d}\omega \qquad (3.3)$$

In equation (3.3), the $e^{i\omega t}$ term is simply a plane wave (like a sine wave) of frequency ω, and the factor $\mathcal{E}(\omega)$ is a factor which determines 'how much' of each frequency ω there is in the overall pulse. $\mathcal{E}(\omega)$ is really just the laser pulse $E(t)$ in the frequency domain. The mathematical machinery that connects these two descriptions is the **Fourier transform**, and there are many excellent resources available if you are interested in the mathematical background of this [4, 5]. We choose to write our pulse like this, because as the pulse propagates through a medium, each frequency in the pulse will be affected differently due to the frequency dependent refractive index $n(\omega)$—so it is much simpler to look at our pulse decomposed into individual frequency components as in equation (3.3). Herein lies the power of the Fourier transform. Problems that are seemingly intractable in the time domain can be rendered trivial in the frequency domain and vice versa. You will encounter this methodology frequently in the physical sciences, so being comfortable converting between the frequency and time domains is a crucial skill.

As the pulse travels through a material, each individual frequency will accumulate a phase, $\phi(\omega)$. If you are not comfortable with this idea, then appendix A provides an introduction to the some of these concepts. The quantity $\phi(\omega)$ is known as the **spectral phase**—that is, the **phase** of each part of the **spectrum** of colours in our pulse. Mathematically, adding the accumulated phase to our waves $e^{i\omega t}$ amounts to simply multiplying them by a phase factor, $e^{i\phi(\omega)}$, such that:

$$e^{i\omega t} \xrightarrow{\text{propagation}} e^{i\omega t} e^{i\phi(\omega)} \tag{3.4}$$

The output pulse $E_{\text{out}}(t)$ can then be written as:

$$E_{\text{out}}(t) = \frac{1}{\sqrt{2\pi}} \int_{-\infty}^{\infty} \mathcal{E}(\omega) e^{i(\omega t + \phi(\omega))} \mathrm{d}\omega \tag{3.5}$$

If we can find the accumulated spectral phase $\phi(\omega)$, then we can calculate the effect that propagation through the dispersive medium has on our pulse. One way to illustrate this effect is to expand $\phi(\omega)$ as a Taylor series around the central frequency ω_0.

$$\phi(\omega) = \phi(\omega_0) + \left.\frac{\partial\phi}{\partial\omega}\right|_{\omega=\omega_0} (\omega - \omega_0) + \frac{1}{2}\left.\frac{\partial^2\phi}{\partial\omega^2}\right|_{\omega=\omega_0} (\omega - \omega_0)^2 + \cdots \tag{3.6}$$

Higher order terms than this can normally be neglected, and we will now discuss each of these terms in turn and gain an understanding of their physical meaning, following notation convention used in Hooker and Webb [1].

Zeroth order term

$\phi(\omega_0) \equiv \phi^{(0)}$. This is the total phase that is accumulated at the central wavelength. This is responsible for the **carrier-envelope phase** (CEP) mentioned earlier, which is interesting but generally only significant for few-cycle pulses, where each pulse only consists of very few optical cycles. This will not be an important factor for the laser pulses we will consider.

First order term

$\frac{\partial\phi}{\partial\omega} \equiv \phi^{(1)}$. This term is called the **group delay**, with units of time, and gives the time taken for the pulse to propagate through the dispersive medium. If the dispersive medium has a length L, we can also intuitively define the **group velocity**, $v_g = L/\phi^{(1)}$. The group velocity defines the *velocity at which the pulse envelope (i.e. the whole pulse) travels*. This is distinct from the **phase velocity**, which is specific to each frequency (each carrier wave) and defines the velocity of the individual carrier waves. Changing the group delay has the effect of changing the temporal position of the pulse, but does not change the shape of the pulse envelope, so does not lead to broadening.

Second order term

$\frac{\partial^2\phi}{\partial\omega} \equiv \phi^{(2)}$. This term is called the **group delay dispersion (GDD)**, and has units of (time)^2. It is the lowest-order term responsible for broadening of pulses via dispersion[3]. It is also generally the main source of dispersion that will broaden our pulses. This term represents the **quadratic phase**—and when it is present it implies that the phase difference between components depends on frequency, so it is no longer possible to achieve the fixed phase relation we would need in order to have

---

[3] One can also calculate the **group velocity dispersion** as GDD/L—the GDD per unit length.

all of our frequencies arrive simultaneously and make a transform-limited pulse. **GDD is the main phenomenon that will cause your pulses to broaden.**

Higher order terms can also play a role, especially **third order dispersion (TOD)**. However, these are much more difficult to compensate for in the lab, and can generally be neglected in most common circumstances.

The above discussion was rather qualitative and a more quantitative description is provided in subsection 3.3.1 for more mathematically-inclined readers. However, if you are happy enough with the qualitative description then the following section can be skipped—it is not necessary to understand the deeper mathematics, but certainly will deepen your appreciation of the underlying physics.

3.3.1 The role of the spectral phase

It is probably not immediately clear why the presence of a *quadratic spectral phase* leads to pulse broadening, whereas *linear spectral phase* does not. We can understand this mathematically by simply plugging equation (3.6) into the exponential term in equation (3.5). Before we do this we will rewrite our equations in terms of a factor $\Delta\omega = \omega - \omega_0$, as this will facilitate the later discussion. We write equation (3.6) as:

$$\phi(\omega) = \phi^{(0)} + \phi^{(1)}\Delta\omega + \frac{1}{2}\phi^{(2)}\Delta\omega^2, \qquad (3.7)$$

neglecting higher order terms and using the contracted notation for the partial derivatives introduced above. The factor $\Delta\omega$ can be simply thought of as a kind of 'bandwidth parameter'—the larger it is, the further we are from ω_0, and the larger the bandwidth. We can then substitute this into the exponential term of equation (3.5) to illustrate the effect of each of the phase terms.

$$\exp[i(\omega t + \phi(\omega))] = \exp\left[i\left((\omega_0 + \Delta\omega)t + \phi^{(0)} + \phi^{(1)}\Delta\omega + \frac{1}{2}\phi^{(2)}\Delta\omega^2\right)\right], \quad (3.8)$$

where we have also made the substitution $\omega = \omega_0 + \Delta\omega$. Equation (3.8) can be further separated by factoring out the term containing ω_0 as the phase terms are all in $\Delta\omega$. We are then left with the following expression for the propagated waves:

$$\exp[i(\omega t + \phi(\omega))] = \exp[i\omega_0 t] \exp\left[i\left(\Delta\omega t + \phi^{(0)} + \phi^{(1)}\Delta\omega + \frac{1}{2}\phi^{(2)}\Delta\omega^2\right)\right] \quad (3.9)$$

To illustrate the effect that each of the phase terms has on the pulse, we will consider them in turn and let the other two phase terms be equal to zero to isolate the effect of each individual term, as in the previous section.

Zeroth order term
Letting $\phi^{(1)} = \phi^{(2)} = 0$ leads to the following expression for the waves after propagation:

$$\exp[i(\omega t + \phi(\omega))] = \exp[i\omega_0 t]\exp[i\Delta\omega t]\exp[i\phi^{(0)}] \tag{3.10}$$

We can clearly see from equation (3.10) that the effect of $\phi^{(0)}$ is to produce a **phase shift** of $\phi^{(0)}$ to the entire pulse. This phase shift is the CEP discussed earlier. This term does not lead to broadening or movement in time.

First order term
Now letting $\phi^{(0)} = \phi^{(2)} = 0$ leads to the following expression for the waves after propagation:

$$\exp[i(\omega t + \phi(\omega))] = \exp[i\omega_0 t]\exp[i\Delta\omega(t + \phi^{(1)})] \tag{3.11}$$

We can see from equation (3.11) that in essence the effect of $\phi^{(1)}$ (the group delay) is to produce a shift to the time axis of our pulse bandwidth, from t to $t + \phi^{(1)}$. The shift is uniform across the whole pulse bandwidth, and so the whole pulse moves in time by the group delay, as expected.

Second order term
Finally, letting $\phi^{(0)} = \phi^{(1)} = 0$ leads to the following expression for the waves after propagation:

$$\exp[i(\omega t + \phi(\omega))] = \exp[i\omega_0 t]\exp\left[i\Delta\omega\left(t + \frac{1}{2}\phi^{(2)}\Delta\omega\right)\right] \tag{3.12}$$

We can see from equation (3.12) that the effect of $\phi^{(2)}$ (the GDD) is to produce a shift to the time axis of our pulse bandwidth, from t to $t + \phi^{(2)}\Delta\omega$. However, this shift is not uniform across the whole pulse bandwidth, as the shift for each individual frequency depends on $\Delta\omega$—how far that frequency is from ω_0. The result of this is that each frequency in the pulse is shifted in time differently, with frequencies further from the central frequency (larger $\Delta\omega$) being shifted further. As each frequency is shifted by a different amount, the pulse broadens, as expected. Moreover, the larger $\Delta\omega$ is, the more severe the broadening—in line with what we expect: broadband pulses undergo more severe broadening than narrowband pulses. This term can also be understood as the rate of change of the group delay with frequency; if the group delay depends frequency then each part of the pulse experiences a different group delay and the pulse broadens.

3.3.2 The form of the spectral phase

We have spoken about the spectral phase $\phi(\omega)$, but it would be nice to have a way to link our quantities $\phi^{(1)}$ and $\phi^{(2)}$ to other known quantities so we can get a feel for how our pulses will broaden and move with time. One way to do this is to note that $\phi(\omega)$ can be written in terms of the wavenumber, $k(\omega)$, as follows [1, 3]:

$$\phi(\omega) = k(\omega)L, \tag{3.13}$$

where L is the distance the light propagates, and $k(\omega)$ is the **wavenumber**. The wavenumber is the magnitude of the **wave vector**, and has units of radians per unit length[4], and points in the direction that the wave propagates. The momentum of the wave is directly proportional to the wave vector. A more detailed examination of the wave vector and other concepts relevant to electromagnetic waves is given in appendix A. The wavenumber can be written as:

$$k(\omega) = \frac{n(\omega)\omega}{c_0},\tag{3.14}$$

where $n(\omega)$ is the frequency dependent refractive index, ω is the frequency of the wave, and c_0 is the speed of light in vacuum. A good way to think of the wave vector is as a vector which propagates in the direction the wave is travelling, and has a magnitude of the wavenumber, which is proportional to the energy of the wave. A wave with a higher frequency has a higher energy, and the units of $k(\omega)$ are essentially inverse wavelength—shorter wavelengths lead to larger $k(\omega)$ and vice versa.

The accumulated phase can be written in terms of the wavenumber as in equation (3.13) because essentially the accumulated phase is the difference in wavenumbers between the wave before and after it has propagated through a medium. So we could write:

$$\phi(\omega) = k(\omega)r_2 - k(\omega)r_1 = k(\omega)(r_2 - r_1) = k(\omega)L,\tag{3.15}$$

where r_1 and r_2 are the positions of the wave before and after propagation, respectively.

We can now learn something about the form of the spectral phase $\phi(\omega)$ by expanding it as a Taylor series around ω_0 again, but now in terms of k rather than directly as ϕ:

$$\phi(\omega) = k(\omega)L = k(\omega_0)L + \frac{\partial k}{\partial \omega}\bigg|_{\omega=\omega_0}(\omega - \omega_0)L + \frac{1}{2}\frac{\partial^2 k}{\partial \omega^2}\bigg|_{\omega=\omega_0}(\omega - \omega_0)^2 L \, d\omega\tag{3.16}$$

We know from before that we can identify these terms as relating to the CEP, group delay, and GDD. The question is now whether we can gain some insight into the physical origins of these quantities by considering the derivatives of the wave vectors. From equation (3.14), we can write the derivatives in equation (3.16) in terms of $n(\omega)$ and find the following.

$$\text{Group Delay} = \frac{\partial k}{\partial \omega} = \frac{n(\omega_0) + \omega_0 n'(\omega_0)}{c_0}\tag{3.17}$$

$$\text{GDD} = \frac{\partial^2 k}{\partial \omega^2} = \frac{2n'(\omega_0) + \omega_0 n''(\omega_0)}{c_0},\tag{3.18}$$

---

[4] This is normally quoted in cm^{-1} for historical reasons.

where the notation $n'(\omega_0)$ and $n''(\omega_0)$ refers to the first and second derivatives of n with respect to ω, respectively. Looking into the equations in equation (3.17) can give us some physical insight into the origins of the group delay and the GDD.

Initially considering the group delay, we can see that the group delay will be larger if the refractive index of the material at ω_0 is large, and also if the rate of change of the refractive index with ω is large. A longer propagation length also produces a larger group delay, intuitively. If the refractive index did not change with frequency (so that $n'(\omega_0) = 0$), we would still have a group delay—the pulse would still be delayed in time as it propagates through the medium.

Now considering the GDD, we see that there are no longer terms that depend only on $n(\omega_0)$. Instead, the GDD depends on both the first and second derivatives of the refractive index with frequency. This means that unless our refractive index is independent of frequency ($n \neq n(\omega)$), then our pulse will **always broaden** as it passes through a material. This is exactly what we described qualitatively earlier in this chapter, where we stated that it was the frequency dependence of n which causes our dispersion. Moreover, we can now go further and say that we will have more GDD (and more broadening) if our refractive index changes more drastically with frequency (so the derivatives are larger), and a situation where $n(\omega)$ is a higher order function of ω will generally cause even more broadening (as then $n''(\omega_0)$ is higher).

3.4 Group delay dispersion

We now return to a more qualitative discussion of the GDD, as needed to enable effective lab working, and to summarise the conclusions of the previous two sections. If you are buying optics and want to assess how much dispersion they will add, you will typically find the GDD quoted in units of fs^2. The effect of the GDD on a pulse can be assessed as follows:

The time $T(\omega)$ required for a given pulse to propagate through a medium can be expressed as a Taylor expansion around a central frequency ω_0 as:

$$T(\omega) = T(\omega_0) + \frac{\partial T}{\partial \omega}\bigg|_{\omega=\omega_0} (\omega - \omega_0) + \cdots \qquad (3.19)$$

$T(\omega_0)$ is just the group delay $\phi^{(1)}$, from our previous definition (the time it takes the pulse envelope to propagate with no dispersion), and $\frac{\partial T}{\partial \omega}$ is the GDD $\phi^{(2)}$. We can then recast this equation as follows (neglecting higher order terms):

$$T(\omega) = \text{Group Delay} + \text{GDD} \times (\omega - \omega_0) \qquad (3.20)$$

So, if GDD is zero, the propagation time is just the group delay, and the pulse is transform limited. However, if the GDD is positive, then the propagation time for frequencies higher than ω_0 is longer than that for frequencies lower than ω_0. The 'blue' end of the pulse takes longer to travel through the medium than the 'red' end; and we have induced **positive chirp** in the pulse. In contrast, negative GDD leads to a **negative chirp**. Also, it is clear that a larger bandwidth (larger difference between ω

and ω_0) will lead to a longer travel time, a more pronounced chirp, and a broader pulse.

Hopefully by now you understand that dispersion (and specifically GDD) causes pulses to broaden: as pulses are dispersed away from the transform limit, they become either positively or negatively chirped, leading to temporal broadening of the pulse.

3.5 Predicting broadening from dispersion

Next we consider how we can predict the broadened pulse length of a Gaussian pulse, which will be of considerable practical utility. Assuming that there is only second order dispersion[5], then the broadened pulse duration τ of an initially transform-limited Gaussian pulse of duration τ_0, having passed through a medium giving a GDD $\phi^{(2)}$ is:

$$\tau = \tau_0 \sqrt{1 + \left(\frac{4 \ln 2 \; \phi^{(2)}}{\tau_0^2} \right)^2}$$

(3.21)

This equation is plotted for a transform-limited 35 fs input pulse as a function of GDD in figure 3.5. Clearly, as the initial pulse was transform limited, then any amount of positive or negative GDD added will cause it to broaden. Also, it can be seen from the equation that a shorter input pulse (smaller τ_0) will lead to a broader output pulse. This is easily explained by the fact that a pulse with a larger bandwidth will experience a wider range of refractive indices, with the result that the different colours in the pulse will spread out more in time.

Figure 3.5 and equation (3.21) both ignore higher order dispersion. This is normally a reasonable approximation, except when considering the shortest pulses. A further point to consider is *how the GDD depends on wavelength*. In general, shorter wavelengths (especially in the UV), experience higher refractive indices, leading to more severe broadening as they accumulate **much** more GDD than an infrared pulse of similar duration. Table 3.2 gives the GVD (simply GDD per unit length) for some common materials at both 800 nm (infrared) and 300 nm (UV).

Table 3.2. GVD for common optical materials at 800nm and 300nm.

Substrate	GVD@800 nm (fs^2 mm^{-1})	GVD@300 nm (fs^2 mm^{-1})
BK7	44.6	206.1
UV-fused silica	36.1	155.6
CaF$_2$	27.7	103.8
Sapphire	58.1	243.1
Air	0.02	0.08

---

[5] This is usually reasonable—and most often if there is third order dispersion we can't do a lot about it anyway.

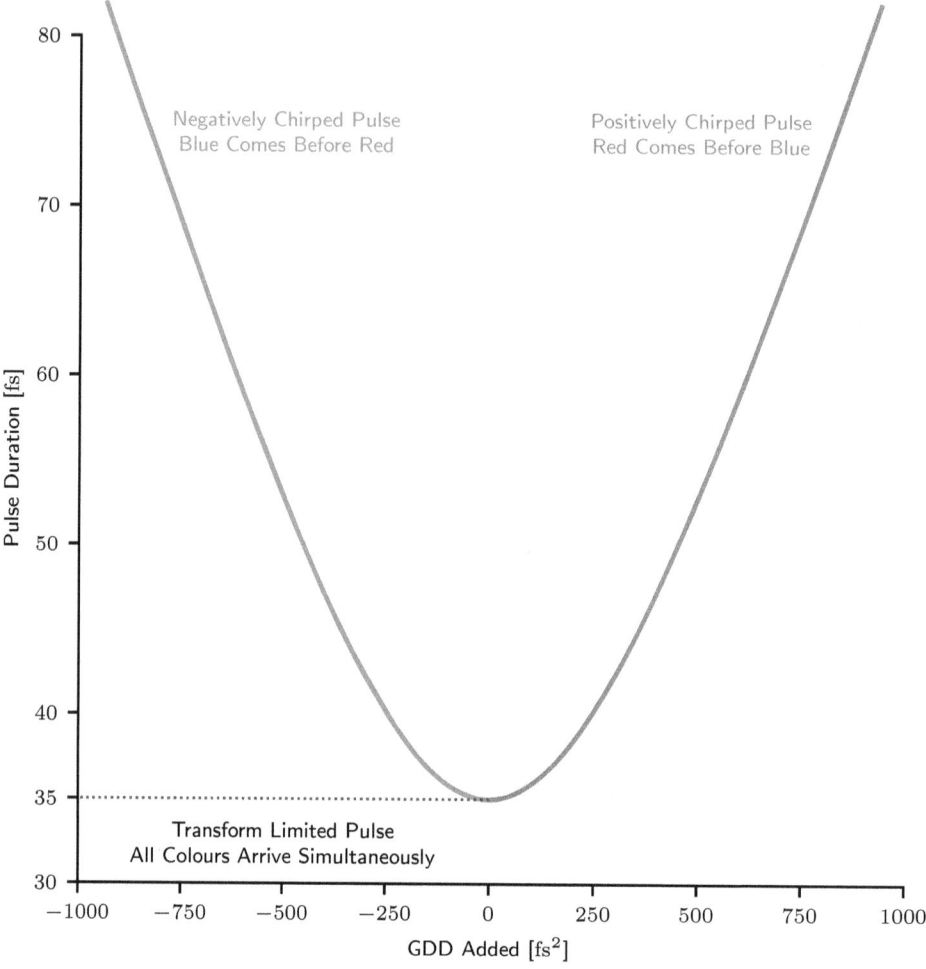

Figure 3.5. Pulse duration of an initially transform-limited 35 fs 800 nm Gaussian pulse as a function of the added GDD. Negative GDD added leads to negative chirp—shown in orange. Positive GDD added leads to positive chirp—shown in blue.

Clearly the amount of dispersion is higher in the UV than in the IR for all of these materials, and passing a 35 fs TL 300 nm pulse through a 5 mm BK7 window will lead to ~1000 fs^2 of added GDD—broadening the pulse from 35 fs to ~90 fs—more than a factor of two! It is therefore important to be aware of the effects of dispersion, especially when working with UV pulses. To illustrate this further, we can consider the actual temporal broadening produced by some typical scenarios, summarised in the table 3.3. The characteristics of the input pulses have been chosen to reflect common outputs of Ti:Sa laser systems.

Table 3.3. Pulse broadening induced by common laboratory scenarios.

Input pulse	Propagates through	Output pulse duration
800 nm, 35 fs, TL	1 cm UVFS	45 fs
800 nm, 35 fs, TL	10 m Air	38 fs
800 nm, 35 fs, TL	10 bounces on ultrafast mirrors	42 fs
800 nm, 150 fs, TL	1 cm UVFS	150.1 fs
266 nm, 70 fs, TL	1 cm UVFS	105 fs
266 nm, 70 fs, TL	10 m Air	80 fs
266 nm, 70 fs, TL	10 bounces on Al mirrors	~70 fs

As should now be clear, to avoid pulse broadening, there are some good rules of thumb:

1. Shorter pulses (broader bandwidth) are a lot more susceptible to broadening than longer pulses (narrower bandwidth).
2. UV pulses are a lot more susceptible to broadening than IR pulses.
3. Minimising transmissive optics is the most important thing to do if you want to keep pulses short.
4. Bouncing on metallic mirrors or ultrafast-coated mirrors is generally OK, but other mirrors can induce quite significant chirp.

3.6 Dispersion of optical elements

Predicting the dispersion associated with a particular optical element is generally a relatively straightforward matter of looking up the value of the GDD or GVD for the optical material in question. Most optics suppliers will give this information readily on their websites. It is also important that you do not **only** consider transmissive optics when thinking about pulse broadening. Mirrors will also contribute some GDD, and especially so if you are not using mirrors that are specially coated for ultrafast pulses. Generally an 'ultrafast' mirror will have a GDD of <30 fs^2 per bounce. The physical origin of GDD from a mirror is discussed further in chapter 7, but is essentially due to different layers of the mirror coating reflecting a different wavelength within the total reflected bandwidth. Therefore, when a pulse bounces on such a mirror, different colours in the pulse travel through subtly different path lengths—leading to dispersion. In a metallic mirror, there are no such layers, and the GDD is (ideally) zero. However, metallic mirrors often have a much lower reflectivity than dielectric mirrors, and are more easily damaged. More detailed discussion of different optical elements (including mirrors), is given in chapter 7.

A very useful resource for finding the GDD for a large number of materials is the Light Conversion Optics Toolbox: http://toolbox.lightcon.com [6]. This contains a huge number of helpful calculators. Under 'Dispersion Calculators' there are applets to calculate arbitrary dispersion, pulse broadening, and much more. These calculators can also include third order dispersion, which can be useful if you need to establish whether or not this can be neglected for your application.

3.7 Pulse compression—compensating for dispersion

Finally, we end this chapter on dispersion by considering how we can remove unwanted dispersion and make our pulses short again. As may be evident from figure 3.5, we want to push our pulse back down the curve towards the transform limit, i.e. towards having zero chirp. How we achieve this depends on whether the pulse has accumulated net positive or negative GDD—whether it is positively or negatively chirped. If we can establish this, then we simply have to add GDD of the opposite sign to shift the pulse back towards the transform limit.

While straightforward in principle, unfortunately there are not really any materials in the UV-near-IR region[6] which exhibit negative GDD. This has two consequences:

1. If our pulses start life as transform-limited pulses, they will almost inevitably end up gaining only positive GDD. The resulting positively chirped pulses will need negative GDD to recompress them back to the transform limit.
2. Adding said negative GDD is challenging due to the lack of materials that exhibit negative GDD in this spectral region.

Given the above, we usually cannot just get another piece of exotic glass and send the pulse through it to recompress it. This can be possible in the mid-IR/far IR, where some common optical materials (such as BK7) add negative GDD[7], but generally is not possible for the 200–800 nm pulses easily produced from Ti:Sa lasers.

Happily, there is a way we can add this negative GDD, by using a **compressor** made using prisms, gratings, or chirped mirrors. These devices are constructed in a geometry such that an initially positively chirped pulse enters (red comes before blue), and within the compressor the red components are delayed relative to the blue components as a result of differing path lengths. The blue components therefore have time to 'catch up' to the red components, adding negative GDD and reversing the positive chirp, compressing the pulse. We will discuss specific compressor geometries further in chapter 5.

References

[1] Hooker S and Webb C 2010 *Laser Physics* 1st edn (Oxford: Oxford University Press)
[2] Milonni P W and Eberly J H 2010 *Laser Physics* 1st edn (New York: Wiley)
[3] Manzoni C and Cerullo G 2016 Design criteria for ultrafast optical parametric amplifiers *J. Opt.* **18** 103501
[4] Kreyszig E 1999 *Advanced Engineering Mathematics* 8th edn (New York: Wiley)
[5] Sanderson G 2018 *But What is the Fourier Transform? A Visual Introduction* www.youtube.com/watch?v=spUNpyF58BY (Accessed: 23-12-2020)
[6] Light Conversion *Optics Toolbox* http://toolbox.lightcon.com (Accessed: 23-12-2020)

---

[6] Which we are interested in when using Ti:Sa lasers.
[7] As these materials have both positive and negative GDD, there is generally a wavelength where they exhibit zero GDD—**the zero dispersion wavelength**.

IOP Publishing

Ultrafast Lasers and Optics for Experimentalists

James David Pickering

Chapter 4

Non-linear optics

Before we consider how to generate, characterise, and practically use ultrashort pulses, it will be useful to discuss **non-linear optics**. Non-linear optics are not exclusively relevant to ultrafast lasers and optics, but ultrashort pulses necessarily have a very high *intensity*, so non-linear effects can be very significant.

The physics and mathematics behind a lot of non-linear optics quickly become quite dense, and would get in the way of the more elementary understanding we need as non-specialists using these pulses in our experiments. For the interested reader, most books on laser physics will offer a more rigorous treatment of non-linear optical effects than will be presented here [1–3]. Here we limit ourselves to brief discussions of the Kerr effect and optical parametric amplification, with a slightly more involved discussion of non-linear frequency mixing.

4.1 Non-linear material response

As discussed previously, a laser pulse can be effectively described as an oscillating electric field (the carrier wave), within a more slowly varying envelope. When a medium (air, glass, anything) is subjected to an electric field E, the response of the material is given by the **polarisation density P**, of the medium. Polarisation density has units of 'dipole moment per unit volume', so if the applied electric field produces a large dipole moment, then the medium is said to be **polarisable**[1]. The polarisation density, P, can be expressed as follows:

$$P = \epsilon_0 \chi^{(1)} E + \epsilon_0 \chi^{(2)} E^2 + \epsilon_0 \chi^{(3)} E^3 + \cdots \tag{4.1}$$

where $\chi^{(n)}$ is the nth order electric susceptibility, and ϵ_0 is the vacuum permittivity. The electric susceptibility $\chi^{(n)}$ is best thought of as the degree to which the material is affected by the field. For weak fields, all terms apart from the first term in equation 4.1

---

[1] Note that this is not the same thing as the polarisation **of** the electric field, which is the orientation of the field oscillation.

4-1

can be considered to be zero. However, for intense fields, such as those we generally have when using ultrafast pulses, the higher order susceptibilities become important. It turns out that in isotropic media (where the medium behaves identically regardless of which of the three symmetry axes we look at—most of the media we use are isotropic), this gives rise to an **intensity dependent refractive index**:

$$n(I) = n_0 + n_2 I(t) \tag{4.2}$$

The first term in this equation, n_0, arises from the first order susceptibility, and is responsible for the normal frequency-dependent refractive index we are familiar with. The second term arises from the *third order* susceptibility[2], and causes the refractive index experienced by a particular frequency within our laser pulse to be different at the most intense parts of the pulse than at the less intense parts. In physical terms, this means that the peak of the pulse in time, and the centre of the beam in space, will experience higher refractive indices than the edges of the pulse/ beam. The extent of this difference depends on the magnitude of n_2; which is material dependent. n_2 is normally positive, so the refractive index is higher for areas of high intensity (such as at the centre of a Gaussian beam, and at the peak of a Gaussian pulse).

This dependence of n on intensity is known as the **Kerr Effect**, and leads to a couple of important phenomena that we need to know a bit about.

4.1.1 Self-focussing and self-phase modulation

We first consider the phenomenon of **self-focussing**. Self-focussing occurs when an intense laser pulse passes through a medium and experiences an intensity dependent refractive index **in space** via the Kerr effect (equation 4.2). As discussed earlier, most laser beams we consider can be well described as **Gaussian beams**, which means that the transverse intensity across the face of the beam is described by a 2D Gaussian function, such that the intensity in the centre of the beam is higher than that at the edges.

If we imagine that a pulse with a Gaussian intensity profile in space passes through a medium, it could (if the intensity is high enough) induce an intensity dependent refractive index in the medium. This effectively leads to a lens being imprinted on the material by the laser beam, and this 'induced lens'[3] leads to focussing of the beam just as a conventional lens would. This focussing of the beam as it passes through the material is known as **self-focussing**, and is illustrated in figure 4.1. Self-focussing is important in the phenomenon of **Kerr-Lens Modelocking**, discussed briefly later.

This process of self-focussing is important for us as laser users, as if we are not careful we can have a situation in which parasitic self-focussing in our optics causes us to inadvertently focus the beam, which can damage or destroy optics that are

---

[2] The second order susceptibility is zero unless the medium is anisotropic—see later in the text.
[3] Sometimes known as **Kerr Lens**.

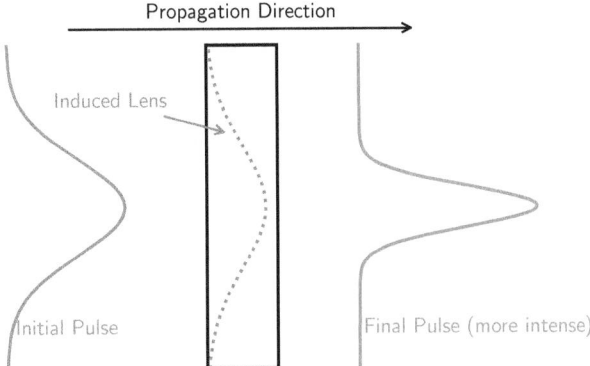

Figure 4.1. Schematic of the self-focussing process. An initial pulse (blue) passes through a dispersive material (black box) and imprints a transient 'lens' on the material via the Kerr effect (grey dashed line). This leads to focussing such that the final pulse (orange) has a higher transverse intensity. The extent of this effect has been exaggerated here for illustrative purposes.

further down the beamline! Aligning optics at low power and then slowly increasing the power with careful observation of the consequences should allow problems to be identified before any damage is caused. You do not want to accidentally focus the beam onto an expensive autocorrelator or inadvertently drill a hole in your vacuum window!

The second effect, self-phase modulation, we will treat more briefly. This is again a result of the Kerr effect, but now due to the intensity at the peak of the pulse **in time** being greater than at the edges. As the intensity of the pulse varies with time (with peak intensity at the peak of the pulse envelope), the pulse will 'see' a time-dependent refractive index as it travels through the medium. This causes a phase shift in different parts of the pulse in time which results in a change in the instantaneous frequency $\omega(t)$ of the pulse which is defined in equation 4.3 [4].

$$\omega(t) = \frac{\mathrm{d}\phi(t)}{\mathrm{d}t}, \tag{4.3}$$

where $\phi(t)$ is the temporal phase of the pulse. Thus, self-phase modulation results in spectral broadening of the pulse, creating new frequencies around the central frequency. This spectral broadening can be important when creation of very short pulses is desired, as effectively you are creating 'extra bandwidth' via the self-phase modulation. Compression of this spectrally broadened pulse will then lead to a shorter pulse than could have been obtained from compression of the initial, non-broadened, pulse. This effect is commonly used in hollow-core fibre broadening, and in **white light generation**.

4.2 Non-linear frequency mixing

One of the most important non-linear effects to be aware of is that of **non-linear frequency mixing**. If we consider mixing of only two photons with frequencies ω_1 and ω_2, then frequency mixing occurs when a third photon ω_3 is produced such that:

Table 4.1. Non-linear frequency mixing schemes.

Input photons	Output photon	Name
ω_1, ω_2	$\omega_1 + \omega_2 = \omega_3$	Sum-frequency generation (SFG)
ω_1, ω_2	$\omega_1 - \omega_2 = \omega_3$	Difference-frequency generation (DFG)
ω_1, ω_1	$\omega_1 + \omega_1 = 2\omega_1$	Second harmonic generation (SHG)
ω_1, ω_2 (where $\omega_2 = 2\omega_1$)	$\omega_1 + \omega_2 = 3\omega_1$	Third harmonic generation (THG)

$$\omega_3 = \omega_1 \pm \omega_2 \tag{4.4}$$

That is, the frequency of the third photon is either the sum or the difference of the frequencies of the two initial photons. In the case where $\omega_3 = \omega_1 + \omega_2$, this is called **sum-frequency generation (SFG)**, and if $\omega_3 = \omega_1 - \omega_2$, then this is called **difference-frequency generation (DFG)**. These are the most general cases, but there are other possibilities with unique names which are summarised in table 4.1. How the non-linear material response described in equation 4.1 can result in the creation of sum or difference frequencies is easily shown with a simple example. We can write an input wave E_ω in terms of the frequency of ω, such as (see appendix A):

$$E_\omega \propto \exp(i\omega t) \tag{4.5}$$

This wave in the non-linear medium gives rise to a second order polarisation $P^{(2)}$ according to equation 4.1:

$$P^{(2)} \propto \epsilon_0 \chi^{(2)} E_\omega^2, \tag{4.6}$$

where:

$$E_\omega^2 \propto \exp(i\omega t)^2 = \exp(i2\omega t) \tag{4.7}$$

Hence, E_ω can produce a second harmonic wave with frequency 2ω through $\chi^{(2)}$, and will do so if the intensity is high enough, **provided that $\chi^{(2)}$ is non-zero**. This is a key concept, and implies that we can only perform this kind of frequency mixing in a medium that lacks inversion symmetry. Most media are isotropic and so have $\chi^{(2)} = 0$, so we need to use an **anisotropic medium**. Generally this is achieved by mixing in an anisotropic non-linear crystal. Given the right crystal, how do we efficiently perform this mixing to make a lot of our second harmonic? For any kind of non-linear frequency mixing, the bottom line is that we have to **combine the photons in a way that conserves momentum and energy**. We will illustrate this using the example of second harmonic generation (SHG).

To combine the photons in a way that conserves momentum and energy, we require:

$$\omega_1 + \omega_2 = \omega_3 \tag{4.8}$$

For energy conservation, since the photon energy E is given by $E = \hbar\omega$. We can write the momentum of a wave in terms of the **wave vector** (see appendix A if this is unfamiliar), k. The photon momentum p is given by $p = \hbar k$, where k is the magnitude of the wave vector $\boldsymbol{k}$. The magnitude of the wave vector in medium i, k_i can be written in terms of the refractive index as follows:

$$k_i = \frac{\omega_i}{v_{p,i}} = \frac{n_i \omega_i}{c}, \tag{4.9}$$

where n_i is the refractive index of frequency ω_i in the medium, and $v_{p,i}$ is the phase velocity of the wave in the medium. The phase velocity can be expressed in terms of the refractive index and the speed of light c, giving the final form shown in equation 4.9.

The momentum conservation criterion can then be written as:

$$\boldsymbol{k}_1 + \boldsymbol{k}_2 = \boldsymbol{k}_3 \tag{4.10}$$

Equation 4.10 is known as the **phase matching condition**. The vector sum of the momenta of the two input photons must be equal to the momentum of the output photon. This is often rewritten in terms of the **wave vector mismatch**:

$$\Delta \boldsymbol{k} = \boldsymbol{k}_3 - \boldsymbol{k}_2 - \boldsymbol{k}_1, \tag{4.11}$$

where the wave vector mismatch $\Delta \boldsymbol{k}$ quantifies the degree of phase matching—in the ideal case, $\Delta \boldsymbol{k} = 0$. If this condition is satisfied, then we can effectively add our input photons together to produce output photons at double the frequency. If we are using collinear beams, we can cancel out the angular dependence of the k-vectors and equation 4.10 reduces to:

$$k_1 + k_2 = k_3, \tag{4.12}$$

where k_1 is the magnitude of $\boldsymbol{k}_1$ and so on. We can now use the relationship between k_i and n_i to write the phase matching condition in terms of refractive indices as:

$$n_1 + n_2 = 2n_3, \tag{4.13}$$

which reduces to:

$$n_\omega = n_{2\omega} \tag{4.14}$$

on substitution of $n_1 = n_2 = n_\omega$ and $n_3 = n_{2\omega}$. Performing the same analysis for generic SFG rather than SHG will show that for SFG, the situation is exactly analogous, but that we require that $n_1 = n_2 = n_3$ for our three photons.

We can get an intuitive feel for the need for phase matching by considering what happens to waves when they are mixed together in a crystal. When equation 4.14 is satisfied, any produced second harmonic travels with the same *phase velocity* as the input fundamental due to them having equal refractive indices. Therefore, any second harmonic photons created by the fundamental at the start of the crystal will be in phase with any second harmonic created by the fundamental at the end of the crystal, and at all points in between. These second harmonic waves then constructively interfere to give the maximum possible second harmonic output, as shown in figure 4.2. This matching of phase velocities is why it is called 'phase matching'.

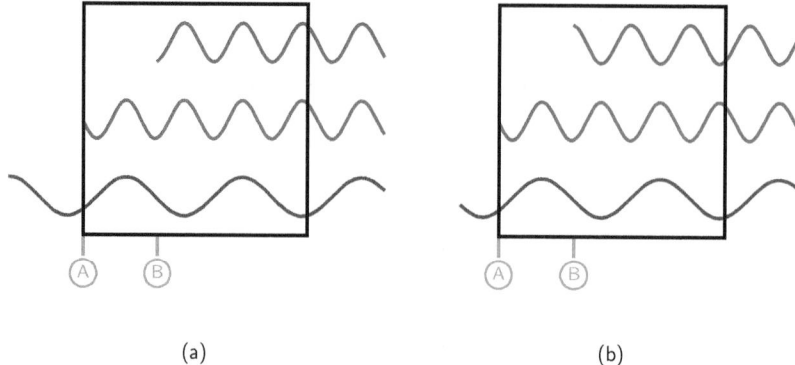

(a)   (b)

Figure 4.2. Simplified illustration of the need for phase matching. In the phase matched case (left) the second harmonic waves (blue) generated at points A and B by the fundamental (dark red) are in phase and constructively interfere, producing intense second harmonic output. In the non-phase matched case (right), the second harmonic wave generated at B is not in phase with that created at A, leading to destructive interference and reduced output. Note that the phase evolution of the fundamental is not presented here.

However, satisfying equation 4.14 is generally not trivial, as we have seen that for most materials the refractive index depends on the frequency! However, there are ways to achieve this, and one way is to use a **birefringent crystal**.

4.2.1 Birefringent phase matching

A crystal that is **birefringent** has a refractive index which depends on the polarisation and propagation direction of the light incident on it. A 3D crystal will have three refractive indices, n_x, n_y, and n_z. If these are all the same then the crystal is **isotropic**, and is not birefringent. If one of these is different from the other two, then the crystal is **uniaxial**. If they are all different, then the crystal is **biaxial**. Birefringence can be exploited to allow us to achieve phase matching, and we will now consider collinear phase matching in uniaxial crystals as an illustrative case.

In a uniaxial crystal, there is one axis which has a different refractive index from the other two, this axis is called the **optical axis** of the crystal, and is shown in dashed blue on figure 4.3. A wave polarised along this axis would experience a refractive index n_e, whereas a wave polarised along another axis would experience a refractive index n_o. If $n_e > n_o$, then the crystal is said to be **positive uniaxial**, and if $n_o > n_e$ it is said to be **negative uniaxial**. A wave propagates through the crystal in a direction defined by the direction of the wave vector, k, shown in orange on figure 4.3. The polarisation plane of this wave is orthogonal to the direction of the wave vector. If the wave is linearly polarised and the polarisation direction is orthogonal to the plane containing k and n_e, then it is known as the **ordinary ray**, with polarisation direction E_{ord}. Conversely, if the polarisation direction lies in the plane containing k and n_e, then it is known as the **extraordinary ray**, with polarisation direction E_{ext}. The ordinary ray always experiences a refractive index of n_o, regardless of the propagation direction. The extraordinary ray experiences a refractive index between n_o and n_e, depending on the angle θ between the wave vector and the OA.

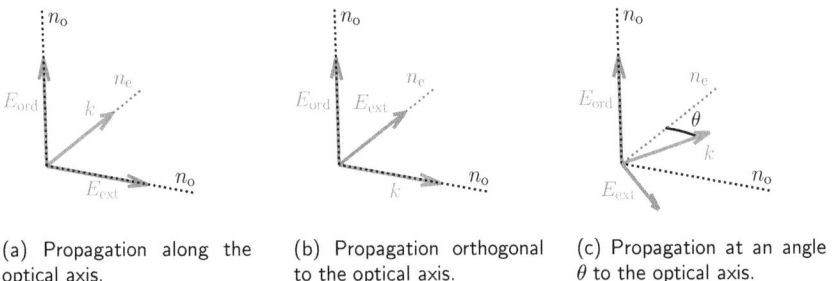

(a) Propagation along the optical axis.

(b) Propagation orthogonal to the optical axis.

(c) Propagation at an angle θ to the optical axis.

Figure 4.3. Propagation of light in a uniaxial crystal with the optical axis (OA—n_e) in dashed blue and the other two axes (n_o) in dashed black. The k-vector (orange) defines the propagation direction, and the light can either be polarised in the plane spanned by the k-vector and the OA (extraordinary polarisation—E_{ext}), or orthogonal to this plane (ordinary polarisation—E_{ord}). Panels (a)–(c) show different propagation directions within the plane spanned by the k-vector and OA.

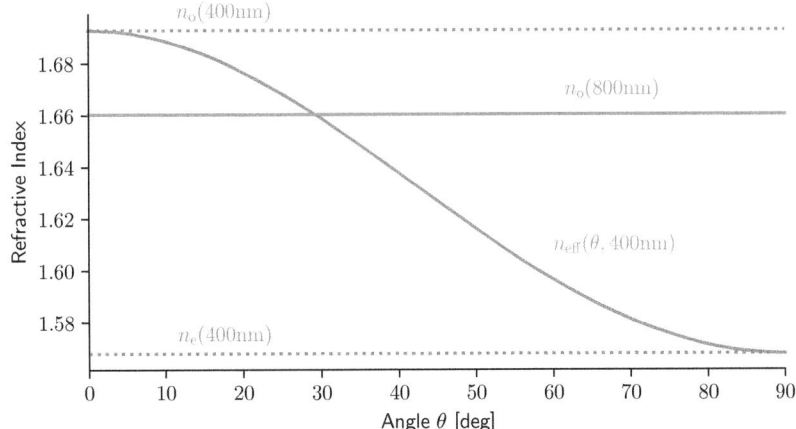

Figure 4.4. Plot of equation 4.15 showing how by tuning the angle θ of a BBO crystal the refractive index n_{eff} (blue) experienced by 400 nm light with extraordinary polarisation can be made to coincide with the refractive index n_o (orange) experienced by the 800 nm fundamental with ordinary polarisation. n_{eff} varies between n_o and n_e for 400 nm light, shown with dashed grey lines. Phase matching would be achieved at the angle where the curves intersect.

Panels (a)–(c) in figure 4.3 illustrate this for the two limiting cases where $\boldsymbol{k}$ coincides with either n_e (a) or n_o (b), and for the case of an arbitrary angle θ (c). The effective refractive index, $n_{eff}(\theta)$ that the extraordinary ray experiences is given by:

$$\frac{1}{n_{eff}(\theta)^2} = \frac{\cos^2(\theta)}{n_o^2} + \frac{\sin^2(\theta)}{n_e^2} \tag{4.15}$$

Equation 4.15 is plotted in figure 4.4 for the case of SHG of 800 nm light in a BBO crystal. To achieve birefringent phase matching, we need to angle our crystal in such a way that the refractive indices experienced by both our fundamental (n_ω) and second harmonic ($n_{2\omega}$) are equal. Figure 4.4 shows that this is possible at an angle of

$\theta = 29.2°$ provided that the fundamental 800 nm has ordinary polarisation and the generated 400 nm has extraordinary polarisation. Turning the crystal to this angle will therefore match the refractive indices and yield efficient SHG.

There are some important points to emphasise regarding this process. Firstly, we have seen that in this case, to achieve phase matching, our output wave must have different polarisation to our input wave. The case where our two input waves (here both 800 nm waves from the same source) have the same polarisation and the output has different polarisation is called **Type-I** phase matching. If the two input waves have different polarisation, it is called **Type-II** phase matching. If all waves have the same polarisation, it is called **Type-0** phase matching. These schemes are summarised in table 4.2. It is not possible to achieve Type-0 phase matching using a birefringent crystal as described here—this requires the use of a technique known as *quasi-phase matching*. Further details on this can be found in references [3, 5].

Another consideration relates to the concept of **phase matching bandwidth**. As we are using broadband ultrashort pulses, it is important that we can achieve phase matching over the whole pulse bandwidth. Generally this will not be problematic if we are only trying to phase match a single colour, but it is a significant consideration if we are looking for a crystal that will allow us to phase match many different colours simultaneously. The mathematics behind this are beyond the scope of this text, but an excellent accessible treatment is given in reference [5]. Ultimately it comes down to the thickness of the non-linear crystal used. A very thick (more than 500 μm) non-linear crystal will have a very high conversion efficiency (allowing production of a lot of second harmonic light), but will also phase match a much narrower bandwidth, and will be very sensitive to the crystal angle. Conversely, a very thin non-linear crystal (less than 500 μm) will be much less sensitive to the crystal angle and phase match a broad bandwidth, but will also have a much lower conversion efficiency and produce a lot less light. Other ways to achieve broadband phase matching involve using non-collinear beams, in which case equation 4.10 turns back into a vector equation, and the angle between the incoming beams becomes an additional parameter that can be adjusted to optimise the characteristics of the output light. Further details and more mathematical treatment of all these concepts can be found in references [2–5], and the Optics Toolbox [6] contains many useful calculators for things like phase matching angles.

Furthermore, angle tuning is not the only way to ensure phase matching in a non-linear crystal. Some crystals have a temperature-dependent refractive index along certain axes, and so the crystal can be housed in an oven and the temperature tuned

Table 4.2. Different polarisation schemes for achieving phase matching. o = ordinary, e = extraordinary.

Name	Wave polarisation (input/input/output)
Type-0	ooo, eee
Type-I	ooe, eeo
Type-II	oeo, oee, eoe, eoo

to match the relevant refractive indices for efficient phase matching [4]. This is often how lithium triborate (LBO) crystals are used in doubling Nd:YLF pump lasers for use in Ti:Sa amplifiers. Performing the phase matching via temperature tuning is referred to as **non-critical phase matching**, in contrast to angle tuning which is referred to as **critical phase matching**. The terminology originates from the fact that in temperature tuning, the relative alignment of the beams and the crystal angles is less sensitive than in critical phase matching.

Finally, a note on buying non-linear crystals. When you buy a crystal, you will see that it is pre-cut for a specific type of frequency mixing—such as SHG or THG at a specified wavelength. This means that when a wave is incident on the flat surface of a crystal, the optical axis is in the right plane such that rotating the crystal in a rotation mount will allow phase matching to be achieved. For example, you could buy a BBO crystal cut at $29.2°$ that is designed for the process depicted in figure 4.4— having the crystal flat to the incoming beam will phase match the process. However, it is not necessary to always use the crystal at the angle it is cut for—you can simply rotate the crystal about the appropriate axis to find the angle needed, within reason.

4.3 Optical parametric amplification

A final non-linear process to mention is that of **optical parametric amplification**. The development of this process has been a boon to many non-specialists (like us) who want to use lots of different colours of light in our experiments, but don't have sufficient practical experience with laser physics to generate them all ourselves. At its core, an optical parametric amplifier (OPA) is just doing the frequency mixing that we have discussed at length in the previous section, but generally over a wider range of frequencies and in a way that is tunable. Using a modern OPA, we can simply type the wavelength of the light we want into a computer, and motors move all the crystals and mirrors around so that the desired wavelengths are produced, which is very convenient. The fundamental idea of an OPA is that a strong **pump beam** with frequency ω_{pump} amplifies a weaker **signal beam** with frequency ω_{signal} via mixing in a non-linear medium. Energy and momentum conservation during this process requires that a third beam, the **idler beam**, is created, with frequency $\omega_{idler} = \omega_{pump} - \omega_{signal}$. This is illustrated in figure 4.5.

The beams are overlapped both temporally and spatially in a non-linear crystal such as BBO to perform the process depicted in figure 4.5, which also needs to be phase matched like the frequency mixing discussed previously. There are some similarities and some differences to the prototypical four-level laser we discussed at length earlier in this book. The similarities are that we still deplete a pump beam whilst amplifying a weaker signal beam, but a crucial difference between an OPA and a four-level laser is that **there is no energy stored within the medium the OPA is performed in**. This is in stark contrast to the four-level laser, in which we were dumping huge energy into a crystal and then extracting part of it and relying on fast relaxation between levels to maintain our population inversion. In the four-level case, we would have had to cool the crystal significantly to prevent damage. In contrast, energy conservation means no energy is actually stored in the non-linear

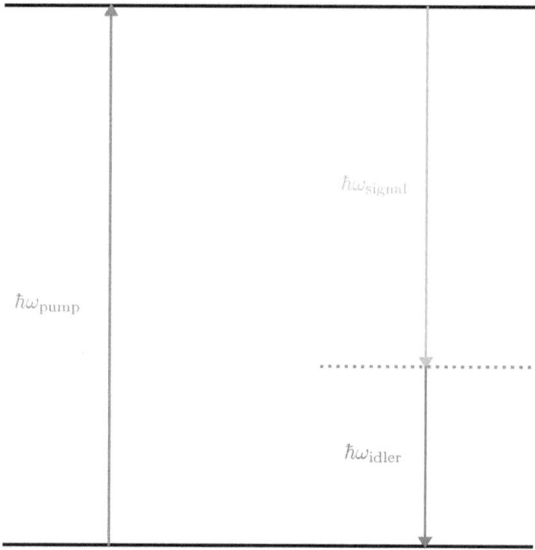

Figure 4.5. Schematic of the optical parametric amplification (OPA) process.

crystal we use for optical parametric amplification—so cooling is not necessary. A useful concept that you may encounter around this subject is that of the **quantum defect**. In the context of lasers, this refers to the difference in energy between the input pump photons, and the output photons from the laser. If the quantum defect is *large*, then the input pump energy is a lot higher than the output energy, and consequently a lot of energy must be stored in (and dissipated from) the crystal. In an OPA, the quantum defect is necessarily zero—due to the energy conservation discussed.

Another key difference is that an OPA can be engineered to have a much broader amplification bandwidth than a typical four-level laser amplifier. This amounts to ensuring there is phase matching over a broad range of bandwidths, which can be achieved by adjusting crystal angles and temporal overlaps within the non-linear crystal in the OPA. This means that if we can produce a weak seed beam that can be tuned to a variety of different colours, we can amplify them all relatively efficiently and have tunable broadband output.

We will discuss briefly how a typical OPA is laid out in chapter 5, but fundamentally we require a few things:

1. A source of a weak signal beam. This is often achieved using **white light generation**, where a beam is focussed into a piece of glass and a broad range of colours is produced via self-phase modulation.
2. An initial amplification stage, where a specific colour from the weak signal beam can be selectively amplified in a non-linear crystal.
3. A power amplification stage, to further amplify this beam up to a power level that is experimentally useful.

All of these stages are generally contained within a single box, and are driven by the output from a femtosecond laser system. OPAs seem like very complex devices, but actually the underlying physics is beautifully straightforward. An accessible introduction to how ultrafast OPAs work (and can be designed) is found in reference [5], and other useful discussions of various concepts can be found in references [1, 3, 4].

References

[1] Hooker S and Webb C 2010 *Laser Physics* 1st edn (Oxford: Oxford University Press)
[2] Milonni P W and Eberly J H 2010 *Laser Physics* 1st edn (New York: Wiley)
[3] Boyd R 2020 *Nonlinear Optics* 4th edn (Amsterdam, Netherlands: Elsevier)
[4] Paschotta R P 2014 *The RP Photonics Encyclopedia* www.rp-photonics.com/encyclopedia.html (Accessed: 22-12-2020)
[5] Manzoni C and Cerullo G 2016 Design criteria for ultrafast optical parametric amplifiers *J. Opt.* **18** 103501
[6] Light Conversion *Optics Toolbox* http://toolbox.lightcon.com (Accessed: 23-12-2020)

IOP Publishing

Ultrafast Lasers and Optics for Experimentalists

James David Pickering

Chapter 5

Generating ultrashort pulses

We will now use the knowledge gained in the preceding chapters to start to gain useful practical knowledge about ultrafast lasers. We start by considering how a laser system produces ultrafast pulses in more detail.

5.1 Laser systems

Note that above we specifically referred to a **laser system** rather than a single laser. This is because ultrafast lasers tend to consist of multiple different lasers that all function together to produce the final ultrashort pulse output. Within your laser system, there will generally be an oscillator, an amplifier, and a compressor. Intuitively, the oscillator produces low-energy pulses, which are amplified by the amplifier, and compressed by the compressor. Within the oscillator and amplifier, there are non-ultrafast pump lasers which pump the gain medium up to a high energy, stretchers to stretch pulses before amplification, and the oscillator may have a separate seed laser too. A schematic of a typical ultrafast laser system, with some representative rep rates and pulse energies for the outputs of each stage is shown in figure 5.1. We will discuss the three main components (oscillators, amplifiers, and compressors) in turn.

5.2 Oscillators

As mentioned in section 1.3, a **laser oscillator** is the part of the laser system where we generate very low-energy ultrashort laser pulses which are amplified further down the road in the laser system. We have already discussed how an oscillator is essentially an amplifier with a cavity built around it to provide feedback, but now we can discuss in more detail the techniques for making an oscillator produce ultrashort (femtosecond) laser pulses specifically.

In chapter 1 we gave a very simple discussion of oscillator construction in terms of the round-trip gains or losses that a particularly lasing pathway experiences in the oscillator cavity. There we said that ensuring that the round-trip gain exceeds the

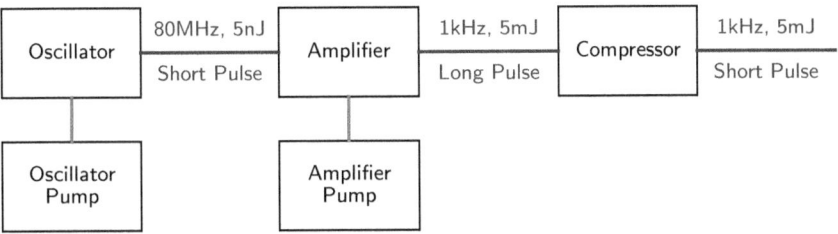

Figure 5.1. Schematic of a typical ultrafast laser system. Pump laser beams are shown in green, and the final ultrashort pulse output beam is shown in red. Representative repetition rates and pulse energies after each stage are shown for illustrative reasons.

round-trip loss was a way to make our cavity oscillate in the desired way. This idea can be put on a more formal footing by considering something called the **quality factor**, or **Q factor** of the cavity. The Q factor is defined as the ratio of the round-trip gain to the round-trip loss, so a cavity with a high Q factor produces a lot of gain, whereas a cavity with a low Q factor is inherently loss.

5.2.1 Q-switching?

We can imagine, then, that a way to produce pulsed laser output is to find a way to *switch* the Q factor of the cavity from low to high for a very short period of time. This is called **Q-switching**, and is a very common way of making a laser oscillator. Q-switching will be familiar if you have ever used YAG lasers, or other nanosecond lasers. At a very basic level, Q-switching amounts to switching the laser on and off very quickly to produce the pulsed output. Energy is built up in the gain medium when the Q is low, and then is quickly dumped out when the Q is high. Electronic switches can easily pulse on for a nanosecond (or so), and so Q-switching can easily produce nanosecond laser output. However, electronic switches are inherently limited in how fast they can switch, and **it would not be possible** to find an electronic switch that can switch fast enough to produce a femtosecond laser pulse. The solution is to use **modelocking** to produce our ultrashort pulses.

5.2.2 Modelocking

We have already met the concept of modelocking briefly in part 1 of this book. To quickly reiterate, there we mentioned how the shortest pulses are made when the largest number of cavity modes all have a fixed phase relationship to one another. When this is the case, the cavity is said to be **modelocked**. This situation was shown in figure 2.4—when colours oscillate in phase, the laser produces a stable train of short pulses. When the phases are random, the laser does not produce a stable train of pulses.

So there are two things to do:
1. Ensure that there is a large enough bandwidth in the cavity to support generation of ultrashort pulses.
2. Ensure that as many of the colours within this bandwidth (the cavity modes) as possible are locked in phase.

The first of these points is easily achieved by using a gain medium like Titanium Sapphire and pumping it to produce broadband laser emission, as described in chapter 1. The second point (ensuring the cavity is modelocked) is harder to achieve, but ultimately it boils down to modulating the cavity gain such that the gain for the desired modelocked process exceeds the gain for the other undesired processes[1]. There are a variety of different methods for achieving this, but they can be broadly divided into **active modelocking** and **passive modelocking** techniques. Active modelocking techniques modulate the cavity gain using some modulator driven by an external signal; whereas passive modelocking techniques rely on the light in the cavity itself to drive a modulator. Active modelocking can produce more stable pulse trains, but as this modulator is typically electronically driven, actively modelocked lasers do not produce the shortest output pulses (although can still produce pulses shorter than 100 fs). Passively modelocked lasers have a modulator driven by the light pulses themselves, so the modulator can be 'switched' much more quickly, and so they produce the shortest pulses (sub 5 fs). More detailed descriptions of modelocking can be found in references [1–5].

There are a very large number of different types of laser oscillator, and the purpose of this book is not to provide an exhaustive list of all the different possibilities (but reference [5] is an excellent place to start if you want this). Rather, we want to understand the basics of how an oscillator works so that understanding the specific oscillator we have in our laser system is not so mystifying. The operating principles of femtosecond laser oscillators are, I think, best shown with an illustrative example: **Kerr-lens modelocking**.

Kerr-lens modelocking
Within passive modelocking, there is one technique that gives an instructive and intuitive illustration of how cavity gains/losses can be modulated by the laser pulses themselves. This is **Kerr-lens modelocking**. We have discussed the creation of a **Kerr lens** in chapter 4; but to briefly reiterate, it is when the intensity of a pulse going into a material is high enough to cause the refractive index to become intensity dependent. This then results in a transient lens being 'imprinted' on the material as the pulse goes through it, as was shown in figure 4.1. This leads to self-focussing, and in a Kerr-lens modelocked laser we exploit this effect to enhance the cavity gain for our modelocked process, while increasing the cavity loss for undesired processes.

Figure 5.2 shows an illustration of one possible scheme for Kerr-lens modelocking[2]. Kerr-lens modelocking exploits the fact that when the oscillator is producing the shortest possible pulses, the intensity of the pulses is necessarily as high as is possible. These pulses imprint a (relatively) strong transient lens on a piece of glass in the cavity called the **Kerr medium**[3]. The pulses then undergo self-focussing to such an extent that

---

[1] The cavity could easily oscillate in continuous-wave mode, rather than being modelocked. This would produce continuous-wave output, which is not what we want when using ultrafast lasers!

[2] There are other geometries, but this illustrates the fundamental principle.

[3] Normally the gain medium itself fulfils this role.

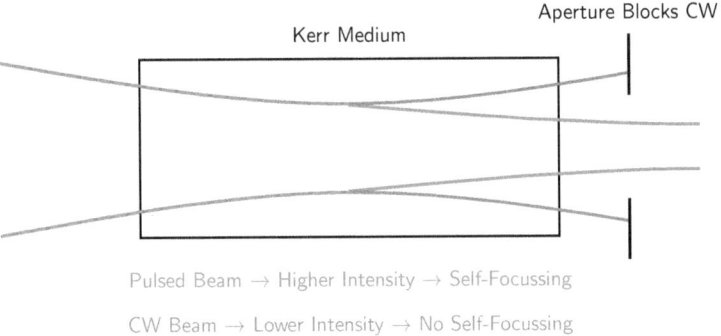

Figure 5.2. Illustration of the operating principle behind Kerr-lens modelocking. Initially co-propagating pulsed (orange) and continuous-wave (blue) beams pass through a thick glass Kerr medium. The less intense continuous-wave beam is blocked by an aperture at the exit of the medium. The more intense pulsed beam undergoes self-focussing via the Kerr effect, and can then pass through the aperture unhindered.

they can pass unhindered through an aperture placed after the Kerr medium. Lasing pathways which do not produce the most intense pulses (or produce cw output) do not undergo self-focussing, and are blocked by the aperture. They therefore experience much higher cavity losses, with the result is that the laser oscillator preferentially produces ultrashort pulses. Another way to do this in practice is by using the nanosecond beam that pumps the gain medium as a pseudo-aperture. The pump beam can be focussed into the crystal such that only the more strongly Kerr lensed beam (the ultrashort pulses) overlap well with the pump beam, so they experience much higher gain. This is more common than use of a physical aperture, but the underlying principle is the same. At this point, the oscillator is modelocked, and a stable train of ultrashort pulses is produced. There will normally also be an intra-cavity pulse compressor to account for the GDD accumulated by the pulse on each round-trip.

We will end our discussion of ultrashort pulse oscillators by emphasising two key points. Firstly, modelocking is inherently a **multimode** phenomenon. The train of ultrashort pulses is created by the interference of a large number of different cavity modes all oscillating in phase. This is in contrast to a Q-switched laser, where the pulses are produced by a sudden reduction in cavity losses, producing a pulse. Secondly, it should be stressed that there are many other different ways to produce a modelocked oscillator than are described here. Researching and understanding what kind of oscillator your specific laser system has is well worth the time spent.

5.3 Amplifiers

We turn now to a discussion of ultrafast laser amplifiers. In chapter 1, we introduced the idea of a laser amplifier as a gain medium that is raised to a high-energy state by a pump laser, and then transfers this energy to a weaker seed laser. The result is that the seed laser beam is amplified and the pump laser beam is depleted. The seed laser in this case is the output from our oscillator, and the pump laser is a high-energy laser beam that can efficiently transfer a lot of energy to our gain medium. We will not discuss pump lasers in more detail, save for saying that they are generally nanosecond lasers, and have a photon energy that matches the gap between the ground state and the pumping bands of whichever gain medium is used. The pump photon energy would be around 532 nm for a Ti:Sa gain medium—this is responsible for the blinding green light that will be familiar to anyone who has (safely) looked inside a running Ti:Sa laser.

There is an additional complication when amplifying ultrashort pulses, which is that an ultrashort pulse delivers its energy in a very short time, meaning that the **peak power** of the pulse is often on the order of 100 TW. Powers this high can easily cause unwanted effects in the gain medium, if they do not damage or destroy it! A solution to this problem is **chirped pulse amplification**, and the Nobel Prize in Physics in 2018 was awarded for the development of this technique [6].

5.3.1 Chirped pulse amplification

Chirped pulse amplification (CPA) is a technique where an ultrashort pulse is **stretched** before it is amplified. Stretching makes the pulse longer, and reduces the peak power back to a level that will not damage the gain medium. We already know that to stretch an ultrashort pulse, the colours need to spread out in time, and so we have to make the pulse chirped. The stretched, chirped pulse is then amplified safely in the gain medium, and the resulting higher-energy amplified pulse is subsequently re-compressed back to an ultrashort pulse. This is operating principle of CPA, and it is illustrated in figure 5.3.

A common question regarding CPA is to ask why the pulses that are intense enough to damage the laser gain medium without being stretched, don't damage mirrors and other optics after the laser output when they are re-compressed. The answer is two-fold. Firstly, reflective optics like mirrors are **much** less susceptible to damage than transmissive optics like a laser crystal, so external mirrors and gratings are much less likely to be damaged. Secondly, the beam inside the gain medium has a very small beam waist, and almost any optic placed in a focussed, compressed, full power beam from the laser output would definitely be damaged! In many ultrafast laser systems producing high-energy pulses, the output beam waist is comparatively large (10 mm or more) to try to minimise this effect.

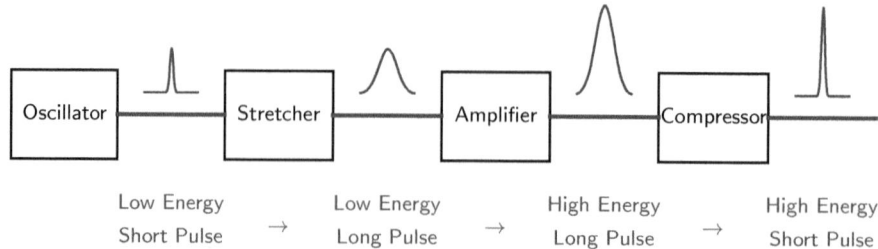

Figure 5.3. Operating principle of chirped pulse amplification. Low-energy ultrashort pulses are stretched, amplified, and re-compressed to produce high-energy ultrashort pulses without damaging the internal optics of the amplifier.

5.3.2 Regenerative amplification

Like laser oscillators, there are many possible ways to build a laser amplifier. We will focus on a design which is commonly found in Ti:Sa lasers, the **regenerative amplifier**. The fundamental operating principle behind regenerative amplifiers is that the laser gain medium is pumped up to a high-energy state using the pump laser, and then the seed pulse passes through the gain medium multiple times to extract the maximum energy possible from the laser medium, achieving the maximum gain possible. So, having pumped the gain medium up to a high-energy state with the pump laser, we need to find a way to make multiple passes through the gain medium. There are two main ways to do this in practice that we will discuss.

Firstly, you could enclose the gain medium inside a resonator, much like how an oscillator is constructed. A simplified schematic of how a regenerative amplifier works is shown in figure 5.4. However, unlike in an oscillator, you do not just let the seed pulse resonate round and round for a long time, as we only want it to complete enough round trips to pass through the gain medium enough times to extract all the energy—in general around 10–20 round trips are required. This requires that the seed pulse is injected at a specific time relative to when the gain medium is pumped, allowed to resonate around the cavity for a fixed number of times, and then coupled out of the cavity after it has extracted all possible energy from the medium. This can be achieved by using an electro-optical gate that can allow you to inject or eject a pulse into a cavity at a specific time via electronic control[4]. Adjusting the timing of the gates can then be used to adjust how much gain you add. This type of amplifier is generally what people refer to when they say 'regenerative amplifier'.

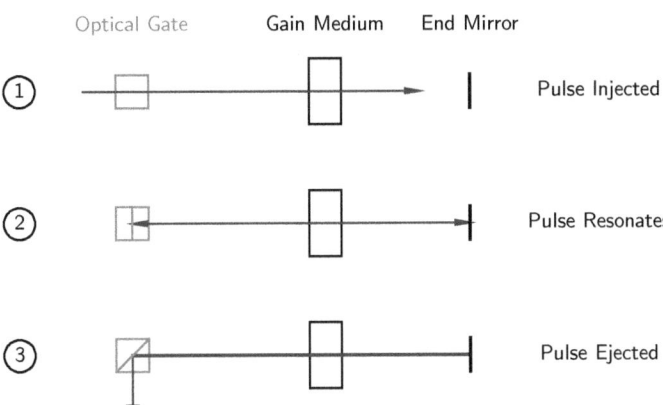

Figure 5.4. Simplified schematic of how a regenerative amplifier works. The electro-optical gate (blue) is opened to allow a pulse to enter (1), then closed such that the pulse resonates around a cavity and passes through the (pumped) gain medium multiple times (2). Finally the gate is re-opened, and the amplified pulse exits the cavity (3).

---

[4] Generally, the cavity round-trip time would be some nanoseconds—so electronically controlled gates would be able to switch fast enough.

A second option is to simply direct the beam using mirrors to make multiple passes through the gain medium, rather than enclosing it in a resonator. Rather than having the beam travel around a 'circular' route and repeatedly enter the gain medium, after each pass through the crystal it is redirected using mirrors make another pass. This process is repeated for the desired number of passes. This type of geometry is commonly called a **multipass amplifier**.

The question remains, though, as to why we would choose one type of geometry over the other? Each type of amplifier has characteristics that are beneficial in certain circumstances. Multipass amplifiers do not need complex driving electronics; but can be difficult to align and generally the same number of passes through the gain medium cannot be achieved as in an electronically gated regenerative amplifier. This means that electro-optical gate driven regenerative amplifiers are able to provide higher overall gain. However, an electro-optical gate contains relatively thick pieces of glass and so the dispersion added by a regenerative cavity is typically much higher than that added by a multipass geometry, making multipass geometries preferable when very short output pulse durations are required[5]. In some laser systems, there is a regenerative preamplifier followed a multipass power amplifier.

5.4 Pulse compression

By the nature of chirped pulse amplification, the pulses that come straight out of our amplifier are very chirped, and therefore very temporally broad. They need to be compressed before they are useful to us as ultrashort pulses—and this is done using a **compressor**. We gave a brief overview of pulse compression in chapter 3, but here we will discuss the geometries of different types of compressor, as well as how to calculate how much GDD they add/remove in more detail.

5.4.1 Compressor geometries

Two very common kinds of pulse compressor are the **prism compressor** and the **grating compressor**. To discuss how each individual type of compressor works, we look at figure 5.5. Initially, the grating compressor (a) works by making the blue components of the pulse travel over a shorter distance than the red components, with the result that negative GDD is added. The prism compressor (b) is similar in principle. However, in this case the red components do **not** have a longer path length, but they travel through more of the prism (as the blue components pass nearer the apex). This means that they experience more material dispersion and slow down relative to the blue components, allowing the blue components to 'catch up'. Grating compressors generally can produce larger amounts of negative GDD than prism compressors—but are more expensive, and generally have a lower transmission. At the time of writing, the best grating compressors are only about 70% efficient (so 10 W uncompressed becomes 7 W compressed). A prism compressor can

---

[5] Whilst a compressor can mitigate the dispersion added by the gate to some extent, it cannot generally compensate for higher-order dispersion. Higher-order dispersion becomes a limiting factor for the shortest pulse durations.

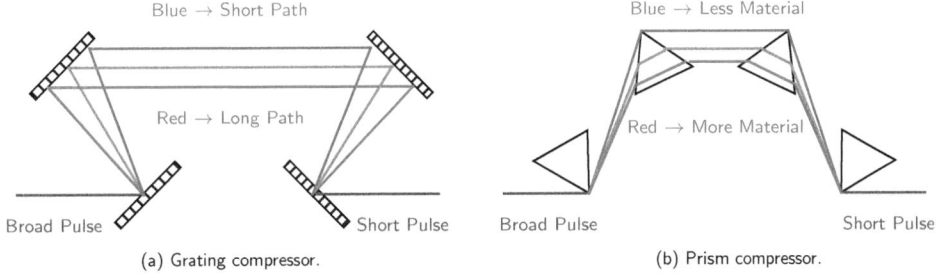

(a) Grating compressor. (b) Prism compressor.

Figure 5.5. (a) Illustration of a grating compressor. Blue components travel over a shorter path length, so 'catch up' to the red components. (b) Illustration of a prism compressor. Blue components travel through less of the prism material than the red components, so they are slowed down less by the prism, and catch up to the red components.

be aligned to have a much higher transmission[6], but often cannot handle very high input powers due to self-focussing (or other non-linear effects). The compressor that you find after a chirped pulse amplifier will almost always be a grating compressor for these reasons. If you plan to build a prism or grating compressor then a good place to start is the Optics Toolbox [7], as here there are tools that let you calculate the GDD (and TOD) for an arbitrary geometry where you can input the relevant parameters. This is ideal as a way to get a feel for how changing the geometry changes the added GDD too.

A third way to produce a compressor is to use a **chirped mirror**. Chirped mirrors will be discussed more in chapter 7, but essentially the mirror is designed such that each bounce on the mirror surface induces negative GDD. Typically, each bounce adds on the order of -50 fs^2. A fixed amount of negative GDD can then be added by simply bouncing on the mirrors multiple times. The usual strategy is then to overshoot, such that your pulses become negatively chirped, and then do fine adjustment using a pair of thin glass wedges (adding positive GDD) to optimise the compression. The advantage of chirped mirrors is that the surfaces can be engineered to effectively control higher-order dispersion over a very wide bandwidth.

There are a few final points to make about these compressors. Firstly, none of the compressors described above can compensate for third order dispersion. One way to do this requires use of a 'GRISM', which is a grating etched into a prism. Another way is to use a **pulse shaper**, which is essentially a device which allows the phase of each individual colour in the pulse to be arbitrarily controlled. Being able to directly and arbitrarily control the spectral phase makes these devices almost the 'ultimate' pulse compressor. Despite their high cost, they are becoming more and more common in university laboratories, especially with recent advances in multi-dimensional spectroscopy. Secondly, the compressors can also be aligned to introduce positive GDD, which can either compensate for negatively chirped pulses, or enable the device to function as a **pulse stretcher** as well as a pulse compressor, but this requires the addition of an extra focussing element in most cases.

---

[6] By placing the prisms at Brewster's angle.

5.4.2 Alternative strategies for compression

Depending on the type of pulses you have, there may also be other ways to compress your pulses. If you have IR pulses (longer wavelength than around 1300 nm), then you may be able to use material dispersion to add negative GDD. We said that generally materials don't tend to provide negative GDD, so we could not easily compensate for the added positive GDD from transmissive optics by sending the beam through a piece of exotic material. While this is true in the visible and near-IR wavelength regions, in the mid to far-IR regions, several common optical substrates actually add negative GDD. For example, BK7 adds negative GDD at wavelengths longer than around 1300 nm. Another useful technique at more conventional visible/near-IR wavelengths is to utilise material dispersion to add positive GDD. If we used a compressor or chirped mirrors to give our pulse a negative chirp, we could then compress it by passing it through some thickness of glass. We could pass it through a thin glass wedge, which can be easily scanned in and out of the beam, changing the amount of material dispersion in a very fine and controllable way.

A final important idea is **pre-compensation** for dispersion. The crux of this idea is that it doesn't matter which order you add the GDD (either positive first or negative first), it all gives the same result in the end. For example, if our ultrashort pulses exit the laser system as transform-limited pulses, then in most cases they will end up only gaining positive GDD as they propagate through our beamline, and so will become positively chirped. A very easy way to compensate for this is to utilise the compressor inside the laser system and alter the chirp of the pulses at the direct laser output such that they are negatively chirped as they exit the laser system. Then, as they accumulate positive GDD travelling through the beamline, they will arrive at our experimental target transform limited. By looking at some kind of intensity dependent experimental signal, the compressor can be adjusted to exactly compensate for the accumulated positive GDD, ensuring that the pulses are as short as possible where it counts: at your experiment.

5.4.3 Optical parametric amplifiers

The final type of amplifier we will discuss is the **optical parametric amplifier** (OPA) that was briefly mentioned in chapter 4. Recall that the key difference with an OPA is that *no energy is stored within the gain medium*. The gain medium is simply a non-linear crystal in which a pump and signal beam with energies $\hbar\omega_{pump}$ and $\hbar\omega_{signal}$ are mixed in a way that amplifies the signal beam, and also produces a third idler beam with energy $\hbar\omega_{idler}$ to conserve energy and momentum. We have already seen an overview of how the process of optical parametric amplification works, so we will now discuss some common ways in which OPAs are constructed. A schematic of how a typical OPA works is shown in figure 5.6.

The pump beam for an OPA is typically the output of another laser system, and the 800 nm output of a Ti:Sa laser system is widely used for this purpose, and is shown in red in figure 5.6. For spectroscopy, the real benefit of an OPA lies in the

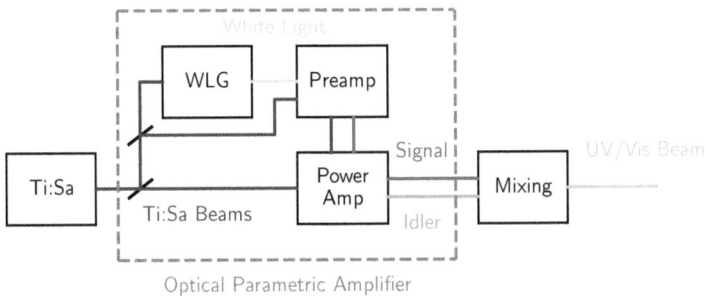

Figure 5.6. A schematic of a typical OPA system. The parts comprising the OPA are surrounded with the dashed grey box, an external frequency mixing stage is also shown at the output. The input Ti:Sa beam is split by beamsplitters shown in black (mirrors in the beamline are not shown).

wide tunability, but this requires that we generate a tunable signal beam from our 800 nm pump beam[7]. A way to achieve this that is commonly seen in OPAs (such as the Light Conversion TOPAS) is to split up the input pump beam, and use a small part of it to drive **white light generation** (WLG).

White light generation is a non-linear optical process where an input beam is focussed into a material (typically a sapphire plate) sufficiently hard that self-phase modulation and other non-linear processes occur. This results in the creation of very spectrally broad 'white light', which is collimated and then directed into an amplification stage. Here the white light beam is overlapped with another small part of the input pump beam in a non-linear crystal. The temporal overlap between the white light and pump pulses, along with the phase-matching angle of the crystal, are then tuned to selectively amplify a specific wavelength. This is how the tunability of an OPA arises.

Like a laser system, an OPA may consist of multiple amplification stages, shown as a preamp and power amp in figure 5.6. The weakly amplified signal beam passes through crystals in each of these stages, gaining energy each time (as each stage is pumped by a fraction of the input pump beam). The phase-matching angles of all the crystals need to be tuned to amplify the desired wavelength of light, and generally an OPA has a large number of motors which are controlled in synchrony to achieve this. The output from the OPA overall is then an amplified signal beam, together with the idler beam generated by the parametric amplification process. For an OPA pumped using 800 nm light, the direct output wavelength is generally tunable between around 1100 nm and 2600 nm. While OPAs do produce this broad range of output wavelengths, they are not uniformly efficient across all wavelengths' bandwidth. An OPA will normally come with **tuning curves**, which are plots of output pulse energy against wavelength, telling you how efficient it is in different spectral regions.

Finally, many commercial OPAs come with an option for **external frequency mixing**—this is essentially a box containing motorised non-linear crystals that can

---

[7] Feeding in a signal beam from an external source would not be possible as it would not be temporally synchronised to the 800 nm pump.

serve to further extend the tuning range. By taking in the signal, idler, and residual input pump beam, non-linear mixing processes can take place to generate many different colours of light. In practice, these extra mixing stages are used to extend the wavelength of an OPA pumped by a Ti:Sa laser either down towards the UV, or further towards the IR. An external frequency mixer generating UV/visible light is shown as the last stage in figure 5.6.

References

[1] Weiner A M 2010 *Ultrafast Optics* 1st edn (New York: Wiley)
[2] Hooker S and Webb C 2010 *Laser Physics* 1st edn (Oxford: Oxford University Press)
[3] Paschotta R P 2014 *The RP Photonics Encyclopedia* www.rp-photonics.com/encyclopedia.html (Accessed: 22-12-2020)
[4] Milonni P W and Eberly J H 2010 *Laser Physics* 1st edn (New York: Wiley)
[5] Paschotta R P 2008 *Field Guide to Laser Pulse Generation* 1st edn (Bellingham, WA: SPIE Press)
[6] Strickland D and Mourou G 1985 Compression of amplified chirped optical pulses *Opt. Commun.* **56** 219–21
[7] Light Conversion *Optics Toolbox* http://toolbox.lightcon.com (Accessed: 23-12-2020)

Chapter 6

Characterising ultrashort pulses

Full characterisation of an ultrashort pulse in a laser beam can be neatly split into three different areas.

- Temporal characterisation.
- Spatial characterisation.
- Energy characterisation.

The second of these three parameters is, strictly, not a characteristic of the *pulse* as much as of the *beam*, but these are the three things to do that will allow the output of your laser to be fully characterised. We will treat each of these areas in turn.

6.1 Temporal characterisation

Knowing how short your ultrashort pulse, measuring the **pulse duration**, is self-evidently an important piece of information to have when working with ultrashort pulses. If you do not know this, you cannot know if your pulse is actually ultrashort! *Temporal* characterisation of the pulse means measuring it in the time domain (measuring the duration); and we will also discuss measuring the arrival times of the different frequencies in the pulse (measuring the chirp).

Measuring the duration of ultrashort pulses is an academic field in its own right, and it will be impossible to give anything approaching a full and rigorous overview here. Instead, we will focus on giving a more practical, qualitative description of various techniques to enable effective laboratory working. I have chosen to discuss two pulse measurement techniques in detail to illustrate how the problem of temporal characterisation is solved, but there are a great many other techniques possible, some of which are listed towards the end of this section for interested readers.

doi:10.1088/978-0-7503-3659-8ch6

6.1.1 Measure what?

Measuring the pulse duration seems like a straightforward concept, but it is useful to define what precisely we need to measure. Ultimately, we would like to measure the full electric field $E(t)$ of our pulse. Note specifically that this is a function of time. $E(t)$ can be conveniently separated into an amplitude, and a term accounting for the phases of the different colours in the pulse. This can be written (in the time domain) as follows:

$$E(t) \propto \sqrt{I(t)} \exp[i(\omega_0 t - \phi(t))], \tag{6.1}$$

where we have replaced the amplitude with the square root of the **intensity profile**, $I(t)$, of the electric field. Note that $I(t)$ is **not** the same thing as the intensity of the laser beam as defined later in this chapter! Later on we will talk about the intensity as the energy per unit area per unit time that our laser beam inflicts on our experimental target. This is generally what people refer to when they say 'intensity', but it is also widely used in physics in the context of equation (6.1). In this instance it is used because in an experiment, we would measure the intensity profile rather than the amplitude directly[1], so taking the square root of the intensity profile gives our amplitude. We will refer to this (time-dependent) intensity as the **intensity profile** going forward, to distinguish it from the intensity of the beam.

The second part of equation (6.1) contains the central frequency ω_0, and the phase $\phi(t)$. So, to fully characterise our pulse, we need to measure the intensity profile, the central frequency, and the phase. Measuring these quantities will be the topic of the remainder of this section.

6.1.2 Measuring the central frequency

Considering the three quantities we wanted to measure, we first consider measuring the **central frequency**. Measuring the central frequency is very simple, and just requires that you have a **spectrometer** that is sensitive to the wavelength of your laser. A spectrometer generally is just a device that can tell you the different frequency/energy components of something that you put into it (such as an ion imaging spectrometer, or an electron spectrometer). In our case, we want an **optical spectrometer**, which is a device that will show you the spectrum of colours contained within a light source. Generally this consists of a box with an optical fibre attached to it that you can point at a piece of paper that your laser beam is hitting. Inside this box are some mirrors and a diffraction grating that separates the colours in the pulse and then images them onto a detector, so the amount of each colour in the pulse can be determined.

When you do this, you should see the spectrum of your laser pulse on the computer attached to the spectrometer. You will then easily be able to read off both the central frequency and the bandwidth of your pulse. You will also be able

---

[1] This is because we would be measuring the square modulus of $E(t)$—multiplying by the complex conjugate so that the effect of the fast oscillating carrier can be separated from the slowly varying amplitude.

to see if the spectrum is the shape you expect. Being able to measure the bandwidth like this also gives you a good way to quickly diagnose problems with the pulse duration. If you suspect that the laser pulses are not as short as they should be, and you then measure the bandwidth and find out it is only 50% of what is expected, then you have a very probable culprit for the cause of your broad pulses.

6.1.3 Electronic measurement?

Before continuing to discuss the different aspects of temporal pulse measurement, we should first address a common question. To anyone familiar with non-ultrafast laser pulses, it may seem odd that we are making such a point of temporal pulse measurement. With a non-ultrafast laser, you can relatively easily set up a photo-diode[2], attach it to an oscilloscope, and read off the pulse duration straight from the oscilloscope. This method works fine, provided that: (a) the time resolution of the photodiode (minimum duration it can measure) is shorter than the pulse; and (b) the time resolution of the oscilloscope is also shorter than the pulse.

Herein lies the problem with electronically measuring an ultrashort pulse. The fastest photodiodes available tend to have a rise time (i.e. how fast it responds—essentially the time resolution) of around 10 ps. Further to this, you would need an oscilloscope capable to measuring this, which would mean an oscilloscope with a bandwidth of 100 GHz of bandwidth. This oscilloscope doesn't exist, and high bandwidth oscilloscopes can be prohibitively expensive anyway. At the time of writing, a 6 GHz oscilloscope retails for around £10 000, so an alternative way to measure pulses is needed.

6.1.4 Measuring the intensity profile: autocorrelation

The problem with electronic pulse measurement is that to measure any pulse, we really need to have an event that occurs on a similar timescale to the pulse. To take a real-world example, if I want to take a clear photograph of something moving quickly, like a sprinter sprinting, then I need a camera where the shutter speed is faster than the timescale on which the sprinter moves. If the camera shutter is open for longer than it takes the sprinter to run through the frame, then the image is blurry. If the camera shutter has opened and shut faster than the sprinter has moved appreciably in the frame, then the image is clear. So, we need to find an event which occurs on a timescale that is at least as fast as our pulse. This is problematic, as ultrashort laser pulses are some of the fastest events in the observable universe! The solution is a neat one, and is to **use the ultrashort pulse to measure itself.**

The fundamental process is one called **autocorrelation**. An autocorrelation fundamentally involves splitting the pulse in half (using, for example, a beamsplitter), and then delaying one half of the pulse relative to the other. The two pulses are then

---

[2] A light-sensitive diode which converts a light pulse into an electrical voltage pulse that can be read by an oscilloscope.

recombined in a way that produces an output signal pulse that **depends on their relative delay**. This recombination is done in a non-linear medium, such as a BBO crystal. Then, one half of our original pulse can be scanned through the other, and the output signal from the recombination will vary as a function of the delay between the two pulses. Depending on the way the pulses were recombined, and how the output signal was measured, we can extract a variety of information about the pulse. A schematic of a generic autocorrelator is shown in figure 6.1.

The diagram in figure 6.1 is generic enough that it will actually suffice for most of our discussion about measuring pulse durations. The key ideas of **using a short pulse to measure a short pulse** and **detecting the signal after mixing in a non-linear medium** are common to the majority of ultrashort pulse measurement techniques. In the case shown in figure 6.1, an input pulse is split and used to measure itself. One half of the pulse is delayed relative to the other, and then the pulses are overlapped spatially in a non-linear crystal. The delay is scanned and when the pulses exactly overlap in time as well as space, then a second harmonic signal (blue) is generated. This signal is measured as a function of the delay between the two pulses. Note that the overlap in the crystal is deliberately non-collinear, so that the second harmonic is produced between the two input beams, and can be easily spatially separated.

The crucial point here is that the output signal can be measured on a *slow detector*, such as a photodiode or spectrometer. This detector doesn't have to have sub-ps time resolution. The temporal information comes from the delay between the two halves of the initial pulse—and we *can* easily delay the two halves of the pulse in

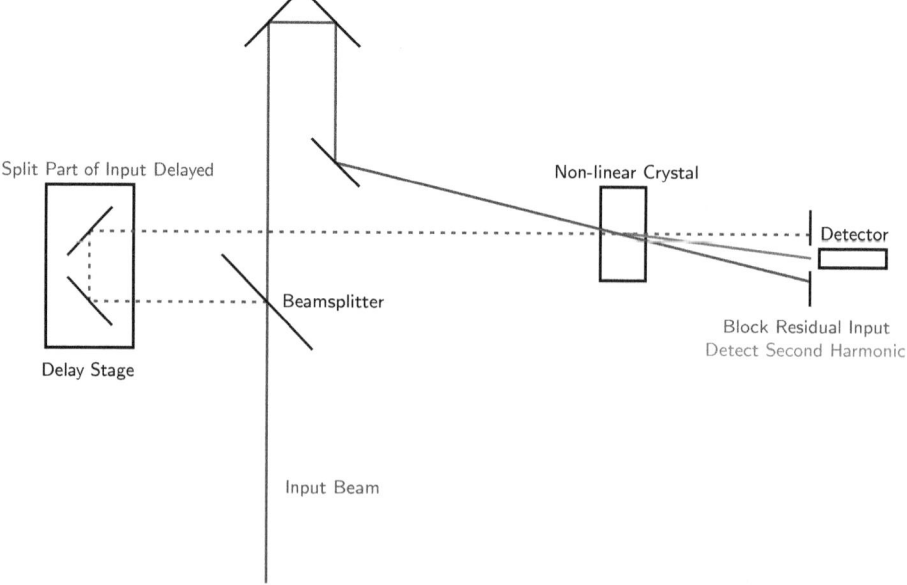

Figure 6.1. A schematic of a generic autocorrelator. Input beams (red) are split on a beamsplitter, and part of the split input (dashed red) is delayed relative to the other half (solid red). The beams are recombined in a non-linear crystal, and the second harmonic (blue) is measured as a function of the delay between the split input pulses.

steps of a few femtoseconds[3]. By exploiting this, we can make our slow diode and oscilloscope or spectrometer measure our ultrafast pulse.

A final note is that autocorrelators generally function in either **scanning mode** or **single-shot mode**; as the name suggests, single-shot autocorrelation measures the entire autocorrelation signal using one laser pulse (one 'shot' from the laser). Scanning autocorrelation relies on the measurement of many pulses in the pulse train to build up the autocorrelation signal. Single-shot measurement removes the impact of any instability or jitter[4] that would impact the accuracy of the autocorrelation measurement. We will now discuss some of the different ways in which autocorrelations can be performed, as well as some of their limitations.

Intensity autocorrelation
The earliest (and arguably simplest) optical autocorrelation performed was an **intensity autocorrelation**, and a schematic of a setup to perform an intensity autocorrelation is shown in figure 6.1, if we used a simple photodiode as a detector. An intensity autocorrelation combines the two halves of a pulse in a non-linear crystal to produce the second harmonic of the input pulse. The efficiency of the SHG process is increased when both halves of the pulse arrive simultaneously (as the intensity is higher), and is reduced when both halves of the pulse arrive at different times. The second harmonic output signal, measured as a function of delay between the two pulses, τ, then has a distinctive intensity profile $I_{\text{Out}}(\tau)$.

$$I_{\text{Out}}(\tau) = \int_{-\infty}^{\infty} I(t)I(t - \tau)\mathrm{d}t \qquad (6.2)$$

The intensity profile of the input pulse $I(t)$ can be determined from $I_{\text{Out}}(t)$, provided that we already know the shape of the input pulse (i.e. if it has a Gaussian, or other known temporal profile). If the input pulse has a Gaussian profile, then the FWHM width of the second harmonic intensity profile is $\sqrt{2}$ times longer than the FWHM width of the input pulse intensity profile. Therefore, by measuring the intensity profile of our second harmonic signal and dividing it by $\sqrt{2}$, we can find out the width of the intensity profile of our input pulse (if we know it is Gaussian).

The width of the intensity profile is useful, and certainly can give you a serviceable number for your pulse duration. However, there are many drawbacks and limitations to this method. Firstly, it tells you the *intensity profile*, and does not measure the **phase** of the pulse, and therefore does not fully resolve the **electric field** of the pulse. Secondly, it requires that you assume the input pulses have a Gaussian (or Lorentzian, or sech[2], or other known) shape. This is quite a drastic assumption, as any noise around the pulse, or satellite peaks, or instabilities in the pulse train can

---

[3] Note that 1 μm of difference in path length (easily attainable with modern delay stages) corresponds to a time delay of around 3 fs.
[4] Jitter specifically refers to when pulses appear in a pulse train in a different place from where they would be expected to be in an ideal pulse train. However, it is commonly misused to just mean 'instability' generally.

mean that the intensity autocorrelation trace is severely distorted[5]. This often leads to dramatic under-estimates of the pulse duration. Nowadays, there are much more powerful and better methods to measure the pulse duration, to the extent that many would say that intensity autocorrelation is an obsolete technique. However, due to the simplicity and relative inexpense of the commercially available intensity autocorrelators they are still widely used, despite their drawbacks. If you only want to get a serviceable number for your pulse duration (and not measure the chirp), and you know your pulse is a stable Gaussian, then intensity autocorrelation works fine.

Phase information?

Throughout the previous discussion we have alluded to the lack of *phase information* in an intensity autocorrelation. There are other kinds of autocorrelation, like the **interferometric autocorrelation**, which do produce some phase information—however, these still are not entirely robust and do not provide a way to unambiguously retrieve the full electric field of the pulse. To do this, we need to measure the **spectral phase** of the pulse—i.e. the phase of each frequency component. Let us briefly recap why this is important.

In an ultrashort pulse, we know that if the spectral phase was a linear function (in time or frequency), then at some point all the colours in the pulse will arrive at the same time, and the pulse is transform limited. If the spectral phase contains higher order contributions, then the different colours in the pulse will never arrive at the same time, and the pulse is chirped and broadened. This was mathematically illustrated in both the time and frequency domains in subsection 3.3.1. Thus, if we know the form of the spectral phase, then we can tell how chirped (or not) our pulse is, and then we can unambiguously measure the pulse duration (and if it is not transform limited, we know what kind of dispersion we need to add to make it transform limited!).

Clearly, being able to measure the spectral phase of our pulse would be very powerful. Knowing the spectral phase would enable us to do several things. Firstly, we could measure the pulse duration *without* having to assume anything about the pulse shape; and secondly we could tell if the pulse is transform limited or not, and if it is not, measure the magnitude and sign of the chirp. This allows us to fully reconstruct the electric field of the pulse from our measurement. It turns out that performing this measurement is as straightforward as simply performing a **spectrally-resolved autocorrelation**, and one way of doing this is with **FROG** measurement.

6.1.5 Measuring the spectral phase: FROG

FROG stands for **Frequency Resolved Optical Gating**, and allows the full electric field of an input laser pulse to be measured. The FROG technique was developed by Rick Trebino [1, 2], and his company 'Swamp Optics' have an excellent website

---

[5] Though single-shot intensity autocorrelations negate some of these drawbacks.

(www.swampoptics.com) that contains not only commercially available pulse measurement/compression equipment, but also a large number of extremely well written and accessible tutorials on the subject of ultrashort pulse measurement [3]. They are presented with a step-up in mathematical complexity compared to the more qualitative descriptions here, so are an ideal next port of call if you want to know more about pulse measurement after reading this chapter.

At its heart, a FROG measurement is simply a spectrally-resolved autocorrelation. An existing autocorrelator can therefore be converted into a FROG by simply placing a spectrometer at the output rather than a photodiode. Our basic autocorrelator in figure 6.1 becomes a FROG if we use a spectrometer as the detector. There are many possible alternative FROG geometries, based on using different non-linear processes for the measurement, which can have advantages in certain situations. These will not be discussed in detail here, but more information can be found in references [1, 2].

Spectrograms and phase retrieval
An obvious question is: 'how does a spectrally-resolved autocorrelation allow the full electric field of the pulse to be measured?' A spectrally-resolved autocorrelation measures a **spectrogram** from which the electric field of the pulse can be retrieved. A spectrogram is a way of representing how the frequency of a pulse varies with time, and is generally shown as a heatmap so that the intensity of various frequency components can be seen as a function of time too. Figure 6.2 shows three simulated spectrograms.

Spectrograms are a beautifully intuitive way to visualise your ultrashort pulse. We met the concept of a spectrogram previously in chapter 3 (figure 3.3). In that case we simply plotted a 1D spectrogram, whereas in figure 6.2 we have plotted the full 2D Gaussian distribution. Considering the leftmost spectrogram in figure 6.2 (unchirped pulse), it is clear that as we travel through the pulse in time (from left to right), then the central frequency of the pulse does not change. This is because the pulse is transform limited. In the central and rightmost spectrograms, clearly the central

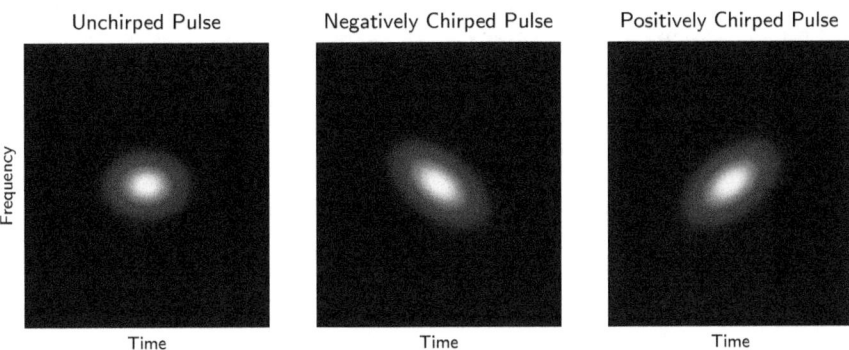

Figure 6.2. Simulated illustrative FROG spectrograms showing (from left to right) an unchirped pulse, a negatively chirped pulse, and a positively chirped pulse. Note that the two chirped pulses are slightly broadened by the chirp. These traces were simulated assuming a PG FROG, see text for details.

frequency of the pulses changes as we travel through the pulse in time, though at $t = 0$ the central frequency is the same as in the unchirped case (and so the 'average' central frequency that would be measured on a spectrometer is the same in all three cases, as the chirp is symmetrical). Note that the rightmost two spectrograms in figure 6.2 would actually look identical in a pure SHG FROG measurement (detecting the second harmonic of the input light), as in this measurement the direction of the time axis is not determined. This direction of time ambiguity can be removed by doing (for example) a polarisation-gated (PG) FROG measurement [2].

However, just having this spectrogram doesn't directly tell us the spectral phase, so we haven't answered the question we set out to answer. The way to extract the spectral phase from the spectrogram is to use a FROG inversion algorithm. The full electric field of the pulse can be reconstructed from the spectrogram using a FROG inversion algorithm. FROG inversion algorithms are widely available, and would come with a FROG if you buy a commercial one, but otherwise are available online. The reason that a FROG spectrogram allows the input pulse to be unambiguously determined is essentially because the sheer quantity of data contained in the pulse spectrogram makes it a quite overdetermined problem—there is a lot more data than are needed to solve the problem. This makes it comparatively easy to reconstruct the full electric field from the FROG spectrogram. More details on how a FROG inversion algorithm actually works can be found in references [1, 4, 5].

Now we have seen two ways to measure the pulse duration: an intensity autocorrelation, and a FROG measurement. The intensity autocorrelation allows us to measure the intensity profile of the pulse electric field, but does not measure the spectral phase. A FROG measurement measures the same things as an intensity autocorrelation, but also measures the spectral phase, so more completely character-ises the pulse. There are also a wide range of different beam geometries for FROG measurements, each with their own advantages and disadvantages, but that enable FROG measurements to measure an extremely wide range of different input pulses. A full review of these is beyond the scope of what we do here, but looking into the tutorials from Swamp Optics [3] is an excellent starting point if you want to know more.

There are many other ways to measure ultrashort pulses besides FROG, and many tend to follow the FROG method of having zoological acronyms. Examples such as Spectral Interferometry for Direct Electric field Reconstruction (SPIDER) [6]; GRating-Eliminated No-nonsense Observation of Ultrafast Incident Laser Light Electric-Fields (GRENOUILLE) [7] both exist. Another notable technique is Multiphoton Intrapulse Interference Phase Scan (MIIPS) [8], which both measures and compensates for pulse dispersion using a pulse shaper. There are many review articles on ultrashort pulse measurement for interested readers [9]. You will find that the acronyms tend to get more forced as time progresses.

6.1.6 One complication: measuring UV pulses

There are some instances where the aforementioned methods for pulse measurement won't work as straightforwardly as I have described here, though the general idea of 'scan two short pulses through each other in a non-linear medium' is still broadly valid. These cases are mostly exceptional cases that you are unlikely to encounter; however, one that you may encounter is when measurement of pulses in the UV region of the spectrum is desired.

The problem with trying to measure a UV pulse with either an intensity autocorrelation or a FROG lies in the non-linear medium. If we were to take a UV pulse, split it in half, and then try to combine the two halves together in a non-linear crystal, we would often find that we're trying to generate light that is either absorbed by the crystal, or simply not detectable by a spectrometer. For example, if we had a 266 nm pulse, using the conventional methods described we would end up trying to double the frequency and make a 133 nm pulse. This would be absorbed by the non-linear crystal (if 266 nm was even within the phase-matching bandwidth), and we would not be able to record our autocorrelation/FROG measurement.

One solution to this is to perform a **cross-correlation** or a **cross-FROG** [10]. In this case, rather than splitting our unknown pulse on a beamsplitter and scanning it through itself, we simply scan a **known reference pulse** through our unknown pulse. The resulting measurement will be a convolution of the unknown pulse and the reference, so provided that we know the duration of our reference pulse, we can extract the duration of our unknown pulse from the measurement. However, it is **very** important that the reference pulse comes from the same laser, as otherwise getting the reference pulse and unknown overlapped in time will be very difficult!

Taking the previous example of measuring a 266 nm pulse as an example, we could mix this in a non-linear crystal with an 800 nm pulse[6]. By selecting the right crystal and turning it to the correct angle for our polarisations, we can perform difference-frequency generation and make 400 nm light when the two pulses overlap in time. Measuring this 400 nm signal with either a diode or a spectrometer will perform a cross-correlation or cross-FROG measurement, which can then be used to extract the pulse duration of the unknown 266 nm pulse. Other ways to measure this pulse could be using either a transient-grating (TG) FROG, or a self-diffraction (SD) FROG [1]. All of these methods have their own advantages and disadvantages—there are many ways to skin a cat in ultrashort pulse measurement, and taking some time to research different methodologies is a good investment.

6.2 Spatial characterisation

We now turn to the question of the **spatial characterisation** of the laser pulse (or laser beam). In section 2.2 we saw that for a Gaussian beam (which is almost certainly what we have), then the quality of the beam was defined using the beam parameter product (BPP), which was given by:

$$\text{BPP} = x_0\theta, \tag{6.3}$$

where x_0 is the *beam waist*, and θ is the *divergence half-angle*. To recap, the beam waist tells us *how big* the laser beam is at a given point, while the divergence

---

[6] As this is probably how you made the 800 nm pulse.

half-angle tells us *how well collimated* the beam is (higher divergence → less well collimated). We can fully[7] characterise our Gaussian beam by measuring these quantities, and can then calculate the quality factor M^2. How we practically can do this is the topic of the following sections.

6.2.1 Beam waist

Let us start the discussion by considering how to measure x_0. This is a logical place to start, as to measure θ requires that we know x_0 in at least two locations anyway (see equation (2.4)); so once we can measure x_0, the rest becomes comparatively more straightforward. Note that a look in the literature will reveal a large number of different ways to define the beam waist, but we talk about the $1/e^2$ width throughout.

How easily you can measure the beam waist depends a bit on how big your beam waist actually is, and how accurately you want to measure it. Are you trying to measure the approximate size of the beam at the laser output to check that the laser is performing to specification? Or are you trying to accurately find out the beam waist of a small focussed beam so you can accurately report the irradiated area of a sample when writing a paper? In the first case, simply taking a card marked with millimetre measurements and looking at the beam on it can do a reasonably good job; but the second case is somewhat more involved, and is what we will focus on for the remainder of this section.

The simplest way to measure the size of a laser beam is to use a **beam-profiling camera**. A beam-profiling camera is just like a webcam that you shine a (heavily attenuated) laser beam on, and the camera software performs a Gaussian fit to the beam spot on the camera. This can give you a fast and accurate readout of the beam waist, and you can even get a 'live' view of the beam so you can monitor the size of the beam as you move a lens back and forth, or change some other optical parameters. Beam-profiling cameras with user-friendly software interfaces are commercially available and lots of companies offer good quality solutions, but it is also perfectly possible to build a beam-profiling camera using an inexpensive commercial webcam, and this has been done in many laboratories the author has seen. A good place to start when doing this is found in references [11, 12], as there are some subtleties to take care of when using a webcam rather than a commercial profiler.

Beam-profiling cameras generally come (or are built with) mountings such that they can be mounted on an XYZ translation stage, so that the camera can be moved in 3D to allow the beam to be easily centred on the sensor. Being able to move the camera along Z (i.e. along the propagation direction of the beam) is also advantageous when measuring a focussed beam, as then you can easily ensure that the sensor is placed at the exact focus by moving the camera along Z to the point where the beam waist is smallest. As a final note, it should go without saying that an intense laser beam can easily destroy a delicate CCD camera chip, so the incident beam will generally require heavy attenuation. Attenuation by around 12 orders of magnitude is typical!

---

[7] At least, as fully as we tend to need to in typical spectroscopy experiments. There are other Gaussian beam parameters like the **Gouy Phase** that can also be calculated, but these are not normally relevant except in special circumstances.

The downside of a beam-profiling camera is that if your beam diameter is very small (typically less than around 35 μm), then the beam won't illuminate enough pixels on the sensor to allow a reliable Gaussian fit to the size. In this case, an option is to use a **scanning-slit beam profiler** or **knife-edge beam profiler**. Both of these designs work by scanning either a small slit, pinhole, or sharp edge through the beam with micrometre precision. Then the transmission behind the slit/edge is measured and recorded. A Gaussian function can then be fit to a plot of the transmission against slit/edge position, and the beam waist easily read from the fit. The measurement takes slightly longer than when using a camera, but it allows smaller beams to be measured.

Building a homemade scanning-slit beam profiler can be a relatively inexpensive and easy way to get a functioning beam profiler set up in a lab. The basic principle is to have a photodiode mounted on an XYZ translation stage, and then to screw a 5 μm or 10 μm slit onto it in a rotation mount. Finding the beam in the slit, and then scanning through both the X and Y directions (by rotating the slit 90°) allows the beam waist in each of these directions to be determined in the way described previously. This method is slow compared to using a beam-profiling camera, and having to move along Z (to find the focus) and measure X and Y at each position can get rather tedious. An even faster way to get a rough idea of the size of a beam (especially useful with unfocussed beams), is simply to use a micrometre stage to scan a razor blade through the beam in front of a power meter. Monitoring the laser power as you scan the blade will allow you to obtain a rough idea of the beam size, but it is possible to burn the razor blade if you do this with a focussed beam!

6.2.2 Divergence

Having measured the beam waist, measuring the divergence half-angle is comparatively straightforward. It simply requires that we measure the beam waist at several different locations along our beam, and then use equation (2.4) to calculate the divergence half-angle. Some commercial beam profilers are designed to be able to do this measurement automatically, so all you have to do is couple the beam into the profiler and it will measure the beam waist and divergence angle, and give you back the BPP and M^2 factor.

If you do not have a commercial instrument that can do this, then the measurement is still relatively simple, but may take some time. How long it takes depends on the Rayleigh range of the beam at the point where you are interested in it. If you want to know the divergence of the beam after focussing, then generally the Rayleigh range will be rather short (on the order of millimetres) and you can easily move your profiling camera/scanning-slit along the propagation direction of the beam using a translation stage. In this case, the measurement is comparatively fast. What can take a lot of time is if you are interested in the divergence of the beam over several metres (for example, to see how well collimated the laser output is). Then you will need to set the profiler up to measure the beam waist in several different locations that could be a metre or so apart. This is not so bad if you are using a profiling camera, but if you need to set up a homemade scanning-pinhole profiler at each location it can quickly get rather tedious!

Thankfully, we often don't actually need to calculate the full M^2 factor, so don't need to calculate the divergence half-angle. Generally speaking in spectroscopy, we are often only really interested in the beam waist at our experimental sample, and measurement of this is sufficient to characterise the laser conditions of our experiment. The author has only calculated the full M^2 factor once—and then only when trying to diagnose a problem with a laser, rather than for an actual experiment.

6.3 Energy characterisation

Having characterised the temporal and spatial shapes of our laser pulse, the final piece of the puzzle is to characterise the **energy** of our pulse. By 'energy' here we specifically mean the energy in the pulse, not the photon energy (wavelength) of the pulse. This quantity is reasonably easy to calculate, and with this information we can then fully characterise our laser pulses by calculating the **intensity**. We will walk through this calculation in the following sections.

6.3.1 Energy and power

To measure the **pulse energy** in your laser beam (the amount of energy within a single pulse), all you need is a power meter and to know the repetition rate of your laser system. The power meter will measure the **average power** (often simply 'power') coming out of your laser (in Watts). The average power is the rate of energy flow from the laser per unit time, and is defined as the energy in a single pulse multiplied by the number of pulses produced per second (the repetition rate). The pulse energy is therefore given by:

$$\text{Pulse Energy (J)} = \frac{\text{Average Power (W)}}{\text{Repetition Rate (Hz)}} \tag{6.4}$$

We can measure the average power using a power meter, and will generally know the repetition rate of our laser system. For example, if a laser operates at 1 kHz (1000 pulses per second), and the average power is 5 W (5 Joules per second), then each pulse contains 5 mJ of energy. 5 mJ may not sound like very much energy, but when that energy all arrives in a few tens of femtoseconds then it can pack a considerable punch! Knowing the amount of energy in a single pulse also allows you to calculate the number of photons in the pulse, provided you know the central wavelength of the laser (the average energy of a single photon). The pulse energy is simply the energy of an individual photon multiplied by the number of photons present. By knowing the repetition rate, you can then easily calculate the number of photons irradiating a target per second—the **photon flux**.

6.3.2 Peak power, fluence, and intensity

In the previous section we showed how to calculate the pulse energy from the measured average power. However, to fully characterise our laser output, we need to account for:

1. The duration of the laser pulse, as the pulse energy arrives in a very short window defined by the pulse duration.
2. The physical size of the laser beam at the point where we want to characterise it, as clearly if the beam is concentrated into a smaller area then it will be more intense.

The first of these points can be accounted for by defining a related quantity to the average power, the **peak power**. Where the average power was defined as the rate of energy flow per unit time the peak power is defined as the rate of energy flow per individual laser pulse:

$$\text{Peak Power (W)} = \mathcal{F} \times \frac{\text{Pulse Energy (J)}}{\text{Pulse Duration (s)}}, \tag{6.5}$$

where $\mathcal{F}$ is a factor that depends on the temporal shape of your pulse, and is around 0.94 for Gaussian pulses. Using equation (6.5) to calculate the peak power then requires that you have measured (or at least have a rough idea of) your pulse duration. If the laser used in the example above produced pulses that were 35 fs long, then the peak power in each pulse is around 140 GW! These peak powers are incredibly high, and this is one of the things that makes femtosecond lasers so useful in the natural sciences.

Looking back to our initial considerations, we can account for the second point by introducing a quantity called the **fluence** of the laser pulse. The fluence is for a Gaussian pulse defined as the energy per unit area:

$$\text{Fluence (J m}^{-2}） = \frac{2 \times \text{Pulse Energy (J)}}{\text{Irradiated Area (m}^2)} \tag{6.6}$$

The factor of two in the numerator arises from the fact that we are measuring a Gaussian pulse [13, 14]. Equation (6.6) requires that you have measured the area of the laser beam at the point of interest (as discussed in section 6.2). If our laser beam is (at least approximately) circular, then generally we can define the irradiated area as simply πx_0^2 where x_0 is the measured beam waist. This then allows the fluence to be defined, and if we take the same example system as previously, and assume that our beam is focussed to a spot with beam waist 35 μm, then the fluence can be calculated to be 2.6 MJ m^{-2}.

Finally, we can go one step further and incorporate both of our initial considerations into one quantity by defining the **intensity**. The intensity is defined in units of power per unit area, and so requires that we know both the pulse duration and the beam waist (or irradiated area). It can then be calculated in a variety of ways, such as the fluence per unit time, or the peak power per unit area. The intensity is generally the number you need when you're writing a paper and want to document the laser conditions, and can be expressed as:

$$\text{Intensity (W m}^{-2}） = \mathcal{F} \times \frac{\text{Fluence (J m}^{-2})}{\text{Pulse Duration (s)}}, \tag{6.7}$$

where the additional factor $\mathcal{F}$ depends on the pulse shape and is 0.94 for a Gaussian pulse, as in equation (6.5). Equation (6.7) gives a straightforward way to calculate the intensity of our laser beam. Useful calculators for a lot of these parameters can be found in the Optics Toolbox [15], and these also give the relevant equations in many cases.

We have now discussed how ultrashort pulses are generated, and how they are fully characterised. Accordingly, we will turn our attention to what happens in between: how do we manipulate our laser pulses so they look and behave as we desire? This requires that you build up a **beamline** from your laser output to your experiment. Doing this is the topic of the remainder of this book.

References

[1] Trebino R 2000 *Frequency-Resolved Optical Gating: The Measurement of Ultrashort Laser Pulses* 1st edn (Berlin: Springer)

[2] Trebino R, DeLong K W, Fittinghoff D N, Sweetser J N, Krumbugel M A and Richman B A 1997 Measuring ultrashort laser pulses in the time-frequency domain using frequency-resolved optical gating *Rev. Sci. Inst.* **68** 3277

[3] Trebino R 2019 *Swamp Optics Tutorials* www.swampoptics.com/tutorials.html (Accessed: 23-12-2020)

[4] Kane D J 1999 Recent progress toward real-time measurement of ultrashort laser pulses *IEEE J. Quantum Electron.* **35** 421–31

[5] Kane D J 2008 Principal components generalized projections: a review *J. Opt. Soc. Am.* B **25** A120–32

[6] Iaconis C and Walmsley I 1998 A spectral phase interferometry for direct electric-field reconstruction of ultrashort optical pulses *Opt. Lett.* **23** 792–4

[7] O'Shea P, Kimmel M, Gu X and Trebino R 2001 Highly simplified device for ultrashort-pulse measurement *Opt. Lett.* **26** 932–4

[8] Lozovoy V V, Pastirk I and Dantus M 2004 Multiphoton intrapulse interference. IV. Ultrashort laser pulse spectral phase characterization and compensation *Opt. Lett.* **29** 775–7

[9] Walmsley I A and Dorrer C 2009 Characterization of ultrashort electromagnetic pulses *Adv. Opt. Photon.* **1** 308–437

[10] Linden S, Kuhl J and Giessen H 1999 Amplitude and phase characterization of weak blue ultrashort pulses by downconversion *Opt. Lett.* **24** 569–71

[11] Langer G, Hochreiner A, Burgholzer P and Berer T 2013 A webcam in Bayer-mode as a light beam profiler for the near infra-red *Opt. Lasers Eng.* **51** 571–5

[12] Keaveney J 2018 Automated translating beam profiler for *in situ* laser beam spot-size and focal position measurements *Rev. Sci. Instrum.* **89** 035114

[13] Newport Optics 2021 *Newport Optics Tutorials* www.newport.com/resourceListing/tutorials (Accessed: 23-12-2020)

[14] Paschotta R P 2014 *The RP Photonics Encyclopedia* www.rp-photonics.com/encyclopedia.html (Accessed: 22-12-2020)

[15] Light Conversion *Optics Toolbox* http://toolbox.lightcon.com (Accessed: 23-12-2020)

Part II

Practical ultrafast optics

IOP Publishing

Ultrafast Lasers and Optics for Experimentalists

James David Pickering

Chapter 7

Optical elements

In the second part of this book we aim to present a guide for construction of optical beamlines. When building up an ultrafast beamline for an experiment, there are a wide range of different **optical elements** (things that can manipulate the beam path or the nature of the laser light) that can be used in the assembly. Coming into a laboratory for the first time and seeing a beamline can be rather daunting, but actually a lot of the individual elements are rather straightforward and just serve to manipulate properties of the light that we have already discussed. This chapter aims to present overviews of the most common types of optical element, together with some simple applications showing how they can be effectively used in the lab.

7.1 General considerations

Almost all optical elements consist of two main parts: the **substrate** and the **coating**. Broadly, the substrate makes up the bulk of the element, providing structural rigidity and providing a surface onto which a coating can be applied.

The substrate can either be an optical element in its own right, manipulating the light directly, or it can simply be the carrier of a coating which manipulates the light. An example of the latter case would be in a metallic mirror, similar to what you would find in a typical household mirror. Here a glass substrate has a metallic coating deposited on it, and this coating is what makes the mirror reflective. Without the metallic coating, the glass substrate is almost entirely non-reflective, and does not work as a mirror. In contrast, an example of the former case would be a lens in a pair of glasses. Here the shape of the glass in the lens produces the focussing behaviour, improving the vision of the wearer. An applied coating doesn't change or alter this primary function, but could enhance other aspects of the performance, such as an anti-glare coating. We will now discuss some important aspects of both substrates and coatings.

7.1.1 Optical substrates

The substrate is the bulk of any optic (physically), and provides it with structural rigidity. The substrate is almost always a kind of optical glass or crystal. In some applications, the choice of substrate is relatively unimportant (the coating does all of the actual useful manipulation!)—but there are some cases where it can be important.

Transmissivity and dispersion
A key area where the choice of substrate is critical is when transmission of a beam **through** an optic is desired. This will happen, for example, if you were focussing a laser beam using a lens, or if you want to separate a beam containing two different colours of light using a dichroic mirror[1]. The transmitted beam has to pass through the substrate, which could be relatively thick (6 mm is common for mirrors). It is therefore important that (a) the substrate does not absorb lots of the light, leading to loss of power; and (b) the substrate does not add a lot of GDD, leading to pulse broadening.

Thankfully this information can be easily found—most manufacturers will show the transmission characteristics of various substrates, and the GDD can be calculated in ways discussed in chapter 3. In the case where avoiding pulse broadening is important, it can be beneficial to use a very thin substrate—these are readily obtainable but substantially more fragile. Table 7.1 lists some common substrates and their transmission characteristics, with common applications.

Clearly, if we want to be sending ultrashort UV pulses through optics (which is generally to be avoided anyway!) then we need to use a UVFS or CaF_2 substrate. However, BK7 is generally fine for wavelengths longer than around 400 nm. You may reasonably ask why we don't just use CaF_2 for everything. The reason is largely that it is more expensive, comparatively fragile, and harder to machine. For these reasons, it only tends to be used where it is truly demanded. Sapphire has relatively poor transmission qualities but a very high damage threshold (to both thermal and mechanical damage) compared to other materials, so is useful in some applications for that reason.

A final point to note is that if you want to send a beam through an optic, you should make sure that the rear face as well as the front face of the optic are both

Table 7.1. Transmission and dispersion performance of common optical substrates.

Substrate	Transmission range (>90% for 10 mm thickness)	Dispersion
BK7	350 nm–2.0 μm	IR: Good—UV: Poor
UV fused silica	200 nm–2.0 μm	IR: Good—UV: Reasonable
CaF_2	180 nm–8.0 μm	IR: Good—UV: Good
Sapphire	Never >90%—reasonable in UV + near IR	IR: Good—UV: Poor

---

[1] A mirror that reflects one colour but allows others to pass through.

polished to optical quality. Some optics are not polished on the rear face (especially mirrors), and so a beam that passes through the coating will simply scatter (diffuse reflection) off the rear face, so the transmitted beam is poor quality (if it exists at all). Conversely—if you do not need to send a beam through an optic, then having a ground rear face (not polished) is safer, as there are fewer reflective surfaces that can result in stray beams flying around the lab.

Damage

Another key property of an optical substrate is in how easily damaged—via mechanical or thermal means—it is. Glass can be scratched—and this will happen in a busy working lab—optics get dropped, and high power laser beams are dumped onto them. The optical substrate needs to be able to deal with all these situations without becoming damaged.

Mechanical damage for our purposes will mostly be asking the question 'how easily can I scratch the substrate?'. The material property that governs this is the **hardness**. Most substrates have reasonably high hardness—but sapphire is exceptionally hard and difficult to scratch. This means it can be made into very thin optics—which can offset the relatively poor transmission properties. BK7 also has a reasonably high hardness (higher than UVFS and substantially higher than CaF_2).

Thermal damage to optics by the incoming laser beams is very common—and will be discussed further in the following section as it mostly damages coatings, rather than substrates. However, of importance in some applications is **thermal expansion** of a substrate via heating by a laser beam. This can cause the size/shape of the mirror to change as it is heated, which can cause the position of beams to drift as the substrate heats up. One way around this is to use bigger beams, and larger optics—so that the heat is distributed over a wider area. Another is to use a specific substrate like **Zerodur**, which has a very low coefficient of thermal expansion, but comes at an additional cost.

Substrate sizing

Optics come in a range of sizes, both in terms of substrate thickness and active surface area (the surface area that the beam is incident on). Standard sizes range from 0.25 inch (6.35 mm) to 2 inch (50.8 mm) in diameter. Optic diameter is inexplicably an area where imperial units still reign supreme (with even metric optical mounts being designed to hold optics with imperial diameters).

Smaller optics are generally used in compact applications, and inside lasers where the geometries are very precise. Larger optics are used for higher power applications and larger beams, such that the power in the beam is spread over a larger area, reducing possibility of damage. In the author's experience, 1 inch diameter optics are a good compromise between cost and ease of use (a larger optic is often easier to align), and seem to be the 'standard' size in many laboratories. It goes without saying that the diameter of the optic needs to be larger than the beam diameter, and having it be around twice as large is often desirable for ease of alignment.

Optical thickness tends to increase with diameter, as the mirror needs to be structurally rigid. Thin optics are desirable for applications where transmission *through* an optic is desired (to minimise power loss and pulse broadening), but this comes at the expense of rigidity. Thin optics are prone to warping if overtightened in an optical mount—the characteristic sign of this is the beam starts to focus unexpectedly, as the optic becomes curved. By not over-tightening thin optics in mounts one can avoid this. Standard thicknesses are normally around 6 mm for 1 inch optics, but thinner optics designed for transmission and reflection of ultrashort pulses will generally be nearer to 3 mm.

Cost

The final property of substrates that is important is their **cost**. Optics get damaged, and will need replacing. You will find that UVFS and BK7 are among the most common substrates for optics, and this is due to a combination of reasonably high damage thresholds and reasonably low cost. CaF_2 or Zerodur substrates are slightly more expensive. Many manufacturers also offer 'economy' mirrors, made from cheaper glass. These may work fine for many applications (or applications where mirrors need frequent replacement), but checking the specifications before buying is worth the time.

7.1.2 Optical coatings

On many optics, the coating is what turns the substrate from a plain piece of glass to a useful optical element. Generally, the flat substrate is extremely well cleaned, and then the coating is deposited via a technique such as electron beam deposition. The coating can affect a multitude of different optical properties, such as reflectivity, dispersion, and polarisation. The most immediately obvious kind of coating is a metallic coating used to create mirrors—as will be familiar from general household mirrors. Metallic coatings can be useful when making mirrors for ultrafast optics, and will be discussed in subsection 7.2.1. A more unfamiliar kind of coating, but much more widespread in ultrafast optics, is **dielectric coating**. Dielectric coatings can produce a very wide range of different optical effects, and so are discussed more at length in later specific sections, with a brief overview of some general points given here.

Dielectric coatings

At its core, a dielectric coating consists of a number of thin layers of material (normally glass or crystal) each with different refractive indices stacked onto a substrate. This construction can create a lot of different behaviours, but the working principle is best illustrated by considering how a reflective dielectric coating for a mirror can be made.

Reflection of incident light will occur off the boundary between the layers that make up the coating[2], but between any given pair of layers the proportion of the reflected light is relatively low. Stacking many layers, and then ensuring that the

---

[2] Reflection will occur off the boundary of any two materials with different refractive indices, but the proportion of reflected light may be very low.

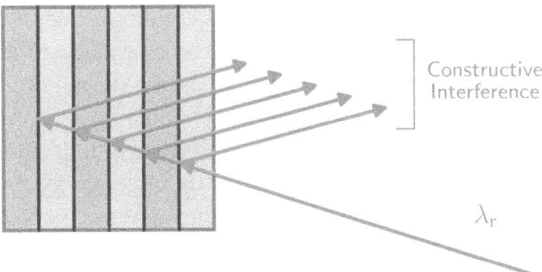

Figure 7.1. Sketch of the operating principle of a high reflectivity dielectric coating. Incoming light of a wavelength λ_r (orange lines) is reflected off boundaries between two materials of different refractive indices (grey and blue shaded). The separation of materials is engineered such that reflections from multiple layers interfere constructively—resulting in high reflectivity.

separation of layers is such that reflections from different layers all end up in phase, increases the reflectivity as the reflections from different layer boundaries constructively interfere. In this way, reflectivities of >99.99% can be achieved. The operating principle is illustrated in figure 7.1.

Figure 7.1 is merely an illustrative sketch of this principle—in reality the layer construction is much more complex to allow a wider range of wavelengths to be effectively reflected (increasing the reflected bandwidth). A very wide range of different coatings can be produced to give almost any specification desired—a look into the catalogues for Layertec will illustrate this!

Another very common dielectric coating is the **anti-reflection coating**, or **AR coating**. This coating is designed in an analogous way to the reflective dielectric coating shown in figure 7.1, except the layers are engineered such that the reflections from different layers end up out of phase with each other and interfere **destructively**, and the coating would not reflect any of the input wavelength. This is the operating principle of the **anti-reflection coating**, which is commonly applied to the surface of transmissive optics to stop them reflecting light of the desired transmission wavelength. This unwanted reflection would cause both power loss, and a safety hazard, as it is another stray beam in the lab that needs to be accounted for.

Manufacturers that produce optical coatings often produce a very wide range of them, covering many possibilities and useful wavelengths. Typically within a specific manufacturer the coatings have a systematic naming convention (such as Thorlabs' 'A' to 'G' anti-reflection coating). These conventions can be useful, as (for example) in a lab using Ti:Sa lasers, generally transmissive optics will need the 'B' coating. Being aware of this can save time when ordering. Different kinds of dielectric coatings with useful specific properties will be discussed in later relevant sections.

Damage and cost
Generally in scientific applications optical coatings are deposited onto the substrate and then left exposed, without a further protective substrate being applied on top (in contrast to say, a household mirror, where the reflective metal surface is enclosed behind glass). This is to avoid extraneous reflections from the additional interfaces

this would create, and in the case of ultrafast optics avoids accumulation of excessive GDD by passing through the front substrate. The downside of leaving the coatings exposed like this is that they are then more easily damaged, much more easily damaged than the substrate itself.

Dielectric coatings are generally much more hard-wearing than metallic coatings, both from a hardness perspective and a thermal damage perspective. All coatings are particularly susceptible to thermal damage if they are not kept clean—any dust or dirt on a coating will absorb energy from an incident beam much more readily, and can quickly result in damage to the coating. Whilst being harder-wearing (and often having superior optical properties, as discussed later in this chapter), dielectric coatings are generally much more expensive than metallic coatings.

7.1.3 Optical labelling

Optics are usually labelled (on boxes) using the acronyms **HR** ('high reflectance'); **HT** ('high transmission'); and **AR** ('anti-reflective'), depending on the coatings applied and intended applications. For example, a mirror designed to separate out 400 nm Ti:Sa second harmonic from the 800 nm fundamental may be marked 'HT800 HR400', showing that it will reflect the 400 nm light and transmit the 800 nm light. Boxes may also show things like the batch number that the coating was produced from (as manufacturers can produce reflectance/GDD curves specific to each batch for demanding applications), or the designed AOI and polarisation. They will also show information such as the substrate size, thickness, and material.

Another aspect of optical labelling is that which is written *on the optic itself*. Sometimes lots of information is laser-engraved onto the edge of the substrate such that even without the original packaging all the necessary info (sometimes including batch numbers) can be found on the edge of the optic. There will also often be an arrow pointing to which side of the substrate the coating is on. This may be laser-etched, but could also be a simple pencil mark, depending on manufacturer. Good practice when taking a new optic out of its packaging for the first time is to check whether or not it has the function of the optic marked on the edge. If it does not, then mark the function on with a pencil, as shown in figure 7.2. Writing the *function* and not the part number is best practice here, as part numbers change, and are usually useless if the manufacturer is not also labelled. An unlabelled, or badly labelled optic is **less use** than a piece of plain glass in a working lab!

7.2 Mirrors

A **mirror**, simply, is something that will reflect any incoming light beam in a **single** direction. Most objects will reflect light to some extent—this is why you can see things—but most objects reflect light in a disordered way, known as **diffuse reflection**. Diffuse reflection is characterised by a well-defined incoming light beam being scattered into many different directions. In contrast, a good mirror will only reflect the incoming light beam in a single direction, which depends on the angle at which the incoming beam hits the mirror (the *angle of incidence*). This is known as a **specular reflection**. The difference is illustrated in figure 7.3.

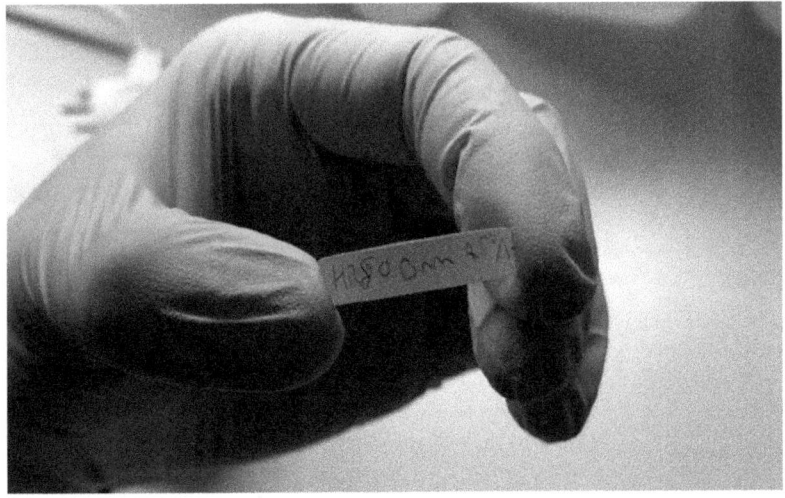

Figure 7.2. A labelled mirror showing clearly the design wavelength, designed angle of incidence, and coated side.

Diffuse Reflection   Specular Reflection

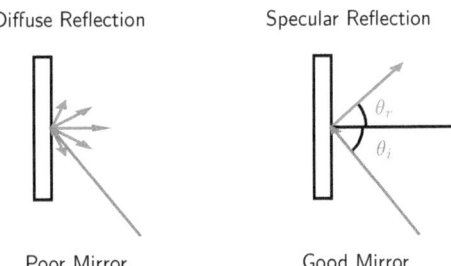

Poor Mirror   Good Mirror

Figure 7.3. Illustration of the difference between specular and diffuse reflection. The incoming beam is shown in blue, and the reflected beam(s) in orange. The angle of incidence θ_i is equal to the angle of reflection θ_r.

So, a good mirror will reflect the incoming light out in a single direction. This is quantified by simply defining the **angle of incidence** (θ_i) and **angle of reflection** (θ_r). These are defined as the angle between the light beam and line orthogonal to the surface of the mirror. The *law of reflection* states that these two angles are the same—so the angle of incidence is equal to the angle of reflection ($\theta_i = \theta_r$). However, the reflectivity of a mirror *can* depend on the angle of incidence, amongst other things such as wavelength and polarisation.

Optical mirrors can be made of a wide variety of materials, which all have different properties. The mirror substrate has some bearing on the performance of the mirror in some applications, as discussed earlier. The mirror coating, however, does the bulk of the 'work' in ensuring that the mirror actually reflects the desired light. Figure 7.4 shows a variety of different optical mirrors with different coatings. We will now discuss different kinds of mirror coating in some detail.

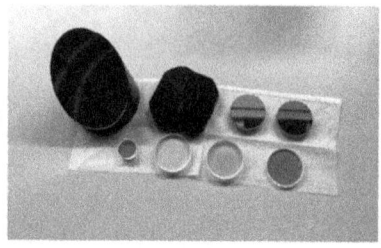

Figure 7.4. Different kinds of commonly encountered mirrors. Clockwise from top left: 2 inch gold parabolic mirror; 1 inch gold parabolic mirror; 1 inch protected silver mirror; 1 inch protected gold mirror; 0.5 inch 635 nm dielectric mirror; 1 inch ultrafast 800 nm mirror; 1 inch broadband dielectric mirror; 1 inch thin harmonic separator. Note that some mirrors look different depending on the angle of incidence the photo is taken at—illustrating that often the reflectivity depends strongly on this.

The mirror **coating** is what turns the substrate from a flat piece of glass into a reflective mirror. There are two main kinds of mirror coatings, **metallic coatings** and **dielectric coatings**.

7.2.1 Metallic mirrors

Metallic mirror coatings will be most familiar, as these are how most mirrors you see in day-to-day life (in bathrooms, for example) are made. The reflective coating is simply a thin layer of metal—the three most common types are aluminium, silver, and gold. Metallic mirrors reflect a wide range of wavelengths[3], and are relatively insensitive to both the angle of incidence and the polarisation of the incoming beam. They also add very little GDD to ultrashort pulses (see next section), so can be used in femtosecond applications. Metallic mirrors are also relatively inexpensive, so can be a cost-effective solution.

However, the metallic coating is relatively soft and is very susceptible to damage. The surface is easily scratched (even when cleaning), and they are more prone to burning than other kinds of coatings. Keeping the mirrors scrupulously clean will minimise the chance of this happening, as burning tends to happen when dirt on the mirror surface absorbs the light and burns, creating a scorch on the mirror. Whilst metallic mirrors have a very large bandwidth, there are differences between the three common metallic coatings, which are roughly summarised in table 7.2.

Within the broad categories of 'aluminium', 'silver', and 'gold', many manufacturers produce subsidiary coatings which are very useful. Chiefly, **UV-enhanced aluminium** is a very efficient and cost-effective way to get a relatively broadband mirror with good reflectivity in the UV region (especially useful with tunable UV sources such as OPAs and dye lasers). UV-enhanced aluminium extends the reflectivity of normal Al mirrors down to around 250 nm and below. **Ultrafast-enhanced silver** is also a very useful coating—as it is engineered to produce very low GDD around the 800 nm region, ideal for use with femtosecond Ti:Sapphire lasers.

---

[3] The **bandwidth** of the mirror is large.

Table 7.2. Metallic mirror coatings and their properties.

Coating	Useful reflectivity range	Comments
Aluminium	400 nm–20 μm	Useful in the UV with additional coating
Silver	500 nm–20 μm	Best in the near IR
Gold	800 nm–20 μm	Best in the mid-far IR

Other coatings designed to enhance reflectivity in other regions, or to improve the damage thresholds, are also available—but note that these overcoatings can also affect other aspects of mirror performance (such as GDD) in a negative way. A quick search of different manufacturers will reveal a wealth of different metallic coatings for many applications.

7.2.2 Dielectric mirrors

Dielectric mirrors are often the 'preferred' choice when compared to metallic mirrors —as they have higher reflectivity and are less easily damaged. They are, however, more expensive and for ultrafast use do not reflect as broad a bandwidth as metallic mirrors[4], and you will find that dielectric mirrors are explicitly marked with the wavelength range they are designed to reflect. Broadband dielectric mirrors tend to have quite poor GDD performance unless they are explicitly marketed as 'ultrafast' or 'low GDD', so it is worth checking documentation before buying. There are also special considerations when using dielectric mirrors with ultrashort pulses.

An ultrashort pulse has a very broad bandwidth, and to reflect the pulse effectively a mirror must reflect the whole bandwidth of the pulse. Some dielectric mirrors are very narrowband, designed for use with narrowband lasers, and these should be avoided for ultrafast applications. Generally you will find mirrors marketed as **'broadband mirrors'** that are designed for this purpose. In addition, the layered construction of the mirror can cause pulse dispersion. Each colour in the pulse will pass through a different amount of the coating when it is reflected—so the different colours will pass through different amounts of material and so the pulse will broaden in time. Dielectric mirrors marketed as **'ultrafast-enhanced'** or **'ultrafast mirrors'** are engineered to minimise/negate this effect, and are well worth the additional cost if you want your femtosecond pulse to still be a femtosecond pulse after reflection! The golden rule is that unless the mirror is explicitly marketed for femtosecond/ultrafast use, assume that it is not suitable.

Dielectric mirrors are also more sensitive to both angle of incidence (AOI) and input polarisation than metallic mirrors. For general lab use, we normally require mirrors that work well at a 45° AOI—resulting in a 90° angle between the incoming and outgoing beams; as well as mirrors that work well at a very small, ~0°, AOI— which reflect the incoming beam back along the input direction. Dielectric mirrors will generally be marked as either 45° or 0° depending on the AOI. 45° mirrors are

---

[4] This can be a benefit and a drawback, sometimes you would like to only reflect a certain colour and dump the rest—a dielectric mirror can be ideal for selectively reflecting the colour you want.

the most common, and generally you will find that mirrors with a 0° AOI are explicitly marketed as **'zero-degree mirrors'**. Regarding input polarisation, you will generally find that in a well-designed 'ultrafast' mirror, the reflectivity in the useful region around the central wavelength does not enormously vary with polarisation. However, as you get further from the designed input wavelength then polarisation becomes more important. The added GDD also depends on the input polarisation, but again only appreciably when you are far from the designed input wavelength. All manufacturers give reflectance and GDD curves for various AOI and polarisations readily online—it is worth checking these before purchasing to check they will work for your application. Manufacturers will also make more data available on request if it is not readily given—such as the transmission of a coating at a wavelength far from its designed reflective wavelength.

7.2.3 Chirped mirrors

A final type of mirror which is specifically useful in ultrafast optics is the **chirped mirror**. This kind of mirror is not typically used for general steering of a laser beam, so was not discussed earlier, but instead finds a niche use in pulse compression.

We saw that a dielectric coating will induce some GDD in a reflected pulse, as different frequency components of the pulse reflect off different 'depths' of the coating, and so travel through different path lengths. In a standard steering mirror, this is generally an unavoidable negative effect that is often negated as far as possible in the coating design. However, in a **chirped mirror**, the coating is engineered to actively induce GDD to aid with pulse compression. A chirped mirror is most often designed to add negative GDD (to counter the positive GDD most ultrafast pulses gain as they propagate), but examples can be found that add positive GDD too.

Chirped mirrors are generally marked by how much GDD they add per reflection ('per bounce'). A typical chirped mirror may add -50 fs^2 of GDD per bounce, so bouncing a beam off a single mirror multiple times can add a relatively large amount of negative GDD which can help to compress the pulse. Some geometries for pulse compression are given in section 5.4.

7.2.4 Cost and damage

Generally speaking, metallic coatings are cheaper than dielectric coatings, but dielectric coatings are **much** more durable. Within dielectric coatings, cost can vary widely depending on the type of coating. As a general rule, coatings which have broader bandwidths, lower GDD, and are designed to reflect a larger number of wavelengths are more expensive than narrowband coatings. Additionally, common coatings are made in bulk and are cheaper—so coatings for use with typical laser output wavelengths (and their harmonics) are often affordable. Companies such as Layertec have extensive catalogues of custom coatings for bespoke applications—but these are often much more expensive (as the cost of starting a production run of a bespoke coating is high). For example, at the time of writing a typical 1 inch 800 nm ultrafast mirror from a standard manufacturer costs around £100. In contrast, a more complex coating (HR 266 nm, 400 nm, and 800 nm with low GDD and >99% reflectivity at all wavelengths) costs closer to £400.

In very high power applications (such as low repetition rate beamlines driving high-harmonic generation), then the power in the beam may be close to the damage threshold of the mirror. To avoid damage, one approach is to expand the beam using curved mirrors such that the power in the beam is spread over a larger area[5]. Sometimes this is not practical, if a specific beam size is desired for a certain application. In this case, it can be more practical to either use a cheap 'economy' mirror that can be easily replaced, or to periodically rotate the mirror in its mounting such that the beam is not incident on one spot for too long.

Application: delay stages
Besides general beam steering, a common specialist use of mirrors is in making a **delay stage**. A delay stage is an assembly that allows the path length of a beam to be changed without changing the pointing direction of the beam. Fundamentally, this consists of two mirrors arranged at 90° to each other such that an incoming beam is reflected back such that the output beam is parallel to the input. The mirrors can then be mounted on a translation stage (electronic or manual) and the position of the two mirrors can be scanned along the direction of the input beam. The amount of delay added to the beam is therefore double the distance by which the stage is moved. A schematic of a simple delay stage is shown in figure 7.5.

Alignment of delay stages can be fiddly and take some time. Ensuring that the two mirrors are placed exactly at 90° to each other simplifies this process enormously. You can do this by hand with careful alignment, but an easier solution is to either buy a dedicated mirror mount that is precisely milled to hold two mirrors at this angle; or to buy a **retroreflector**, which is a single optical element that contains two reflective surfaces at precisely 90° to each other. Once you have the 90° angle, it is

Input Beam

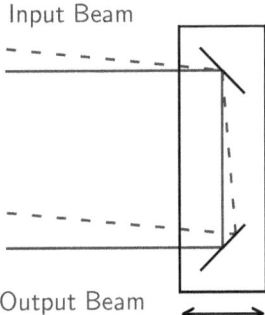

Output Beam

Figure 7.5. Schematic of a delay stage. The mirrors in the black box are movable in the direction indicated by the double-headed arrow. The solid red line shows a well-aligned input beam. The dashed red line shows a poorly-aligned input beam—in this case the pointing direction of the beam will not stay the same as the stage is moved.

---

[5] The enormously high energy beams at the National Ignition Facility in the USA are reflected off very large (~50 cm^2!) mirrors.

then a case of ensuring that the input beam is coming in exactly straight. This is shown with the solid red line in figure 7.5, and when this is satisfied, the mirrors can be translated on the stage without moving the pointing direction of the output beam. If the input beam is coming in at an angle, shown with the dashed line in figure 7.5, then the output beam will also be angled, and moving the stage will change the pointing direction of the beam.

The best way to ensure the stage is 'running true' is to couple the beam onto the stage, and then look at the output beam a long distance (several metres or more) away. If the position of the output beam does not move as the stage is scanned, then it is well aligned. This alignment is particularly important if you are scanning the delay of beams focussed down to small focal spot sizes. It is critical to place irises around the input beam once you have it coupled onto the stage effectively, so you can easily get back to good alignment if things drift.

7.3 Beamsplitters

A **beamsplitter** is an optical element which does exactly what you expect it to. When a beam is incident on a beamsplitter, part of the beam is reflected off the surface of the beamsplitter, and the rest of the beam is transmitted through it—like a partially reflective mirror.

7.3.1 Standard beamsplitters

The most simple kind of beamsplitter will split light in a predictable ratio. The ratio of the reflected to transmitted light is called the **splitting ratio**, and is commonly quoted as numbers such as '50:50' or '30:70'. This functions as would be expected: sending 2 W of light onto a 50:50 beamsplitter will result in 1 W being reflected and 1 W being transmitted. Generally a beamsplitter will have a front surface that is coated to give partial reflectivity, and a rear surface which is AR coated. This makes sure that the only reflection you get is from the front surface, rather than there being two reflections from the front and back surface.

The most standard type of beamsplitter has a splitting ratio that is independent of both the polarisation state of the incoming beam, and the colour of the incoming beam[6]. In contrast, a **polarising beamsplitter** is one that will reflect one polarisation state (s or p) whilst transmitting the other, these will be discussed further in section 7.4.

Application: splitting laser outputs
Standard beamsplitters are commonly used directly in front of laser outputs, when it is known that the output of one laser system needs to be shared in some fixed proportion between multiple end users. A common example of this is when the output of a laser system is used to pump an external OPA, or other frequency conversion stage. Such non-linear frequency conversion stages are very sensitive to input power, so having fixed beamsplitters directing the beams to them prevents the

---

[6] Provided that the colour is within the bandwidth that the coating reflects.

input power being erroneously adjusted up or down. The only drift possible in this scenario is if the laser system output power drifts, and this would be quickly spotted and corrected in a working lab.

7.3.2 Dichroic beamsplitters/mirrors

An often important type of beamsplitter for people doing spectroscopy is the **dichroic beamsplitter**. A dichroic beamsplitter is one that will reflect one colour of light whilst transmitting another, and it is really another word for a dichroic mirror. There are many different types, but of particular use are ones designed to separate harmonics of a laser output from the fundamental. These are often marketed as 'harmonic separators', and are designed to (for example) separate the 400 nm Ti:Sa second harmonic from the 800 nm fundamental. Harmonic separators are available in a very wide range of configurations to suit all kinds of common laser applications. The key things to make sure of when buying dichroic beamsplitters for an ultrafast application are firstly that the coating has a broad enough bandwidth (as usual); and secondly that the substrate is thin enough to not broaden the transmitted pulse extensively.

Application: separating harmonics from fundamental
Following non-linear frequency conversion, such as second harmonic generation, the beam will contain light of both the desired harmonic and of the input fundamental (as the conversion is generally at most around 40% efficient). A convenient way to separate out the harmonic is to reflect it off a series of dichroic beamsplitters/mirrors. This is illustrated for a simple case in figure 7.6, but can easily be extended to more complex systems in an analogous way.

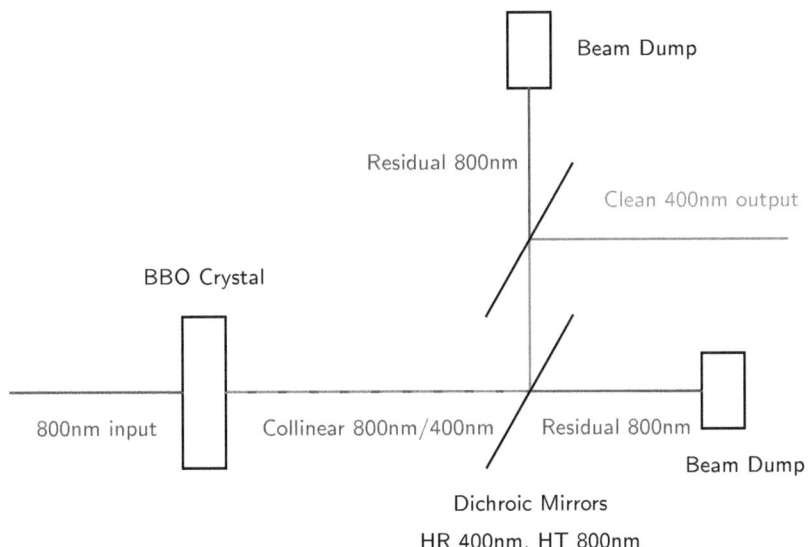

Figure 7.6. Illustration of how dichroic mirrors can be used to separate 400 nm frequency doubled light from the 800 nm fundamental. Multiple mirrors may be needed to fully remove all the residual fundamental, which should be safely blocked in beam dumps.

7.4 Polarisation optics

Polarisation optics is a catch-all term that refers to any kind of optical element that can vary the polarisation of a laser beam. We have already mentioned one common type of polarisation optics, the polarising beamsplitter, in the previous section. We can divide our discussion of polarisation optics into two parts. Firstly, optical elements which will allow different polarisation states to be separated out from a beam (e.g. only allowing s-polarised to be transmitted and reflecting/absorbing p-polarised light), these are **polarisers**—elements that can turn an incident unpolarised beam into an output polarised beam. Secondly, optical elements which allow the polarisation state of a beam to be actively changed, such as turning an s-polarised beam into a p-polarised beam. These are **waveplates**. The difference between a polariser and a waveplate may not be immediately obvious; ultimately a polariser only *separates out* existing polarisation states from a beam, and does not actually *change* the polarisation state. A waveplate actively *rotates* the polarisation state of a beam (or can change it from linear to circular polarisation, for example). This is illustrated in figure 7.7.

An accessible mathematical framework for understanding polarisation in optics uses the **Jones Calculus**, where the polarisation state of light is expressed using a vector (the *Jones Vector*), and each optical element that can affect polarisation is written as a transfer matrix (the *Jones Matrix*). The effect of a number of different optical elements on some initial input light can then be simply calculated using matrix algebra. This method is described in detail in many standard textbooks on optical physics, together with another formalism using the **Stokes Parameters** [1, 2]. As usual, such detail is omitted here, in favour of a qualitative overview and more discussion of some considerations that apply specifically to ultrafast lasers.

7.4.1 Polarisers

Polarisers work by transmitting one polarisation state of a beam and reflecting or absorbing the other, thus only 'letting through' one polarisation state. The effectiveness of a polariser is determined by its **extinction ratio**, which is defined as the ratio of the transmission of the desired polarisation state to the transmission of the undesired polarisation state. Very high-end polarisers can achieve extinction

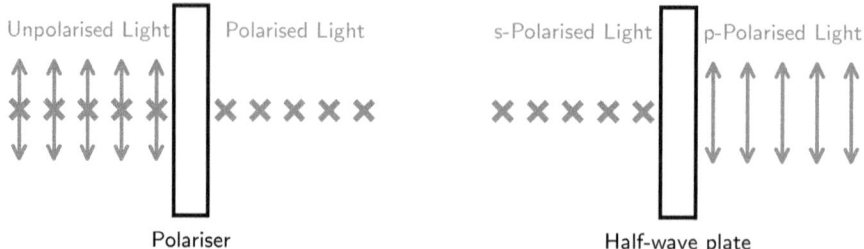

Figure 7.7. Illustration of the function of a polariser (left), as opposed to a half-wave plate (right). The polariser acts to produce well-polarised light from an unpolarised beam, whereas the waveplate actively rotates the polarisation to a different direction.

ratios as high as 100 000:1 or more. When you buy a polariser, it will have an indication of the **transmission axis**[7]. Light that is polarised parallel to the transmission axis will pass straight through the polariser—it will transmit light that is polarised parallel to the transmission axis. The remainder of the light is either reflected out of the polariser (more common) or absorbed by the polariser (less common). In most applications, polarisers that reflect the non-transmitted polarisation are more appropriate, as then the reflected beam can be used or dumped safely, rather than simply heating up the polariser.

There is a very wide variety of different kinds of polariser, as any look through a manufacturer's catalogue will show you. A thorough review of all the different types of polariser is beyond the scope of what we do here, but there are three situations where a polariser is commonly needed in ultrafast laser experiments which will be discussed in some detail. Some initial remarks about choosing and using polarisers are given below.

In an experiment using ultrafast laser pulses, adding polarisers can be dangerous as a lot of polarisers consist of very thick pieces of glass which can add lots of unwanted dispersion to our input pulse, broadening it significantly. One approach to mitigate this is to use a *thin-film polariser* (TFP). A thin-film polariser has a dielectric coating which is engineered to (normally) reflect s-polarised light and transmit p-polarised light. This coating is engineered using interference effects in the same way that a dielectric mirror coating is made. However, TFPs do not generally work over a very wide wavelength range. The range will be specified by the manufacturer, but it is not uncommon to find TFPs that only work well at one single wavelength (or a narrow range of wavelengths). In ultrafast optics, this is especially important, as a TFP marked as an 800 nm TFP may look ideal for use with a Ti:Sa ultrafast laser, but actually may not work well over a broad enough bandwidth to polarise the entire pulse well. TFPs marketed as 'broadband' or 'ultrafast' are designed specifically for this purpose, and worth the additional cost. TFPs are also commonly supplied with AR coatings, which can be helpful.

Another important type of polariser that you may encounter is the **cube polariser**. These polarisers contain very thick glass, so can add a lot of GDD and are not normally ideal for ultrafast use. However, they transmit a very broad bandwidth, and are good for diagnostic use in areas where the added GDD is unimportant. **Crystal polarisers** such as Glan-Taylor or Glan-Laser polarisers can produce the highest extinction ratios, but also contain relatively thick pieces of material, so tend not to be used with ultrafast pulses (except for diagnostics) for GDD reasons.

Polarisers are generally very sensitive to the angle of incidence of the input beam. The specific angle depends on the specific polariser, but generally the flat face of the polariser needs to be placed either at 90° or at Brewster's angle[8] to the propagation direction of the incoming beam.

---

[7] Either directly marked on the mounting (if the polariser came mounted), or supplied with the accompanying literature.

[8] Brewster's angle is the angle of incidence where p-polarised light is totally transmitted by a material. It varies from material to material but is normally in between 50 and 60 degrees.

Application: polarising a beam

The simplest use of a polariser is, literally, to use it to polarise an existing beam! The output from your laser system is probably (in theory) polarised, but may actually not be incredibly well polarised if you were to measure the polarisation[9]. Some experiments are exquisitely sensitive to input polarisation, so it is desirable to 'clean up' the polarisation before using the beam in the experiment. A simple way to do this is to insert a thin-film polariser into the beam path somewhere before the beam reaches the region of interest.

By taking the portion of the beam that is reflected by the polariser, we can clean up the polarisation whilst not adding any significant GDD to our pulse. If p-polarised light is desired for the experiment, then adding a waveplate (see below) into the reflected beam path will allow the polarisation to be varied. Alternatively, it may be possible that the substrate of the polariser is thin enough to not incur any significant dispersion (this can be easily calculated for your application using the optical toolbox), and in this case taking the transmitted portion will also work well.

7.4.2 Waveplates

A **waveplate** is an optic that actively changes the polarisation state of the light incident on it. There are two common types of waveplates, a **half-wave plate** (HWP), and a **quarter-wave plate** (QWP), and these are discussed in more detail below. Waveplates are sometimes referred to as 'retarders' or 'phase retarders'—a name which derives from the mechanism by which they function, where one component of the polarisation of a wave is delayed or 'retarded' relative to the other by some fixed amount. For a HWP, this is $\frac{\lambda}{2}$, whereas for a QWP it is $\frac{\lambda}{4}$, where λ is the wavelength of the incident light. This corresponds to a phase shift of 180° (in the HWP case), or 90° (in the QWP case) in the delayed component relative to the other. As a result of the delay being a fraction of the incident wavelength, it should be intuitive that waveplates generally do not work over a very wide range of input wavelengths. When trying to rotate the polarisation of a broadband ultrashort pulse, this can be problematic, so generally we use **zero-order waveplates** (see below) to mitigate this effect.

Waveplates are made of a birefringent material, so that the refractive index of light along one of the axes of the waveplate is different to that along another, producing the aforementioned phase shift. These axes are commonly called the 'fast axis' and 'slow axis', referring to the speed of light polarised parallel to the axis. An introduction to how waveplates modify polarisation states is given in section 7.4.2, but many standard textbooks will give a more detailed and mathematical treatment [1, 2].

---

[9] This can be easily done by measuring the transmission of your beam through a polariser using a power meter. If (for example) the polariser transmits 99 mW when the transmission axis is vertical, and 1 mW when the transmission axis is horizontal, then the light is 99% polarised—provided that you know the light is linearly polarised.

When buying waveplates you will find that there is a distinction between lower cost 'multiple-order' waveplates, and higher cost 'zero-order' waveplates. The difference can be illustrated by considering two kinds of HWP. A zero-order HWP would produce a phase shift which is exactly half of the wavelength of the incident light. In contrast, a multiple-order HWP would produce a phase shift which is an integer number of wavelengths, plus half a wavelength. This difference arises due to the construction of the waveplates. A zero-order waveplate is constructed from two very thin, highly polished, pieces of material that are sandwiched together. In contrast, a multiple-order waveplate is made from a single piece of comparatively thick material, so is cheaper and easier to produce. However, multiple-order waveplates are generally more sensitive to the wavelength of the incident light than zero-order waveplates, and if the incident light is far from the designed wavelength then the performance of the waveplate can decrease dramatically (the imparted phase shift becomes further from the ideal value, so your HWP might become a QWP when it is used far from the design wavelength).

We know that when using ultrafast lasers, the bandwidth of the laser pulse is generally quite broad, and we want the entire pulse to be affected by the waveplate as equally as possible. Put another way, you do not want the polarisation components of the blue part of your pulse to experience a different phase shift from the components in the red part, as then the blue part of the pulse could end up in a different polarisation state than the red part! For this reason, zero-order waveplates are generally preferred for ultrafast work, and specific zero-order achromatic waveplates are also available for extremely broadband applications. These have exceptionally flat retardation across a very broad bandwidth, but come at an increased cost compared to standard zero-order waveplates.

Waveplates and polarisation
We have discussed briefly the effect waveplates have on light, but have not discussed how this can actually influence the polarisation of our laser beam in the lab. To gain an initial insight into how this works, we will consider the action of a typical half-wave plate (HWP). HWPs are commonly used as **polarisation rotators**, because linearly-polarised light can be rotated through an arbitrary angle using a HWP. To understand how this works, first remember that a waveplate is made of a birefringent material, so has two axes with distinct refractive indices: the fast axis and slow axis.

Now consider what would happen if linearly-polarised light was aligned exactly along the fast or slow axis of a HWP[10]. In this case, nothing would happen! As all of the light would experience the same refractive index, so no phase shift would be observed. However, if the polarisation of the light was aligned somewhere between the fast and slow axis, then the situation is more complex.

Figure 7.8 shows the case that the input polarisation of the light (lower green arrow) makes an angle θ to the fast axis of a HWP, which is placed at the origin in

---

[10] So that the electric field is oscillating exactly along the axis.

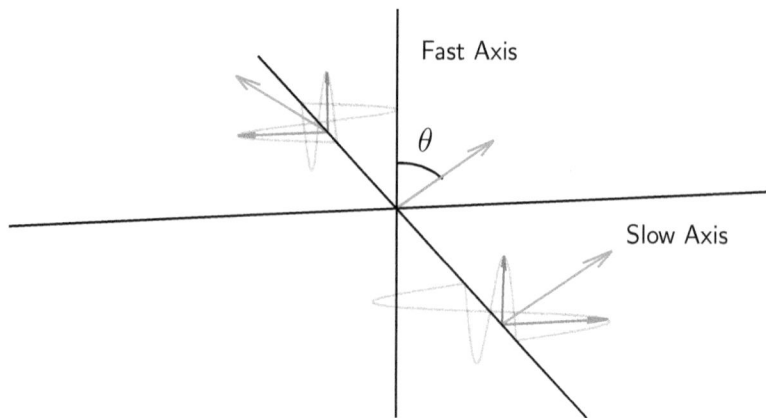

Figure 7.8. How a HWP can rotate polarisation. Linearly-polarised light travels from the lower right, through a HWP at the origin, to the upper left. The overall polarisation of the light is shown in green, with the fast and slow polarisation components in blue and orange, respectively. The net effect is rotation of the polarisation by 2θ, where θ is the angle between the incident polarisation and the fast axis.

figure 7.8. We could decompose the overall input polarisation state into two components, one parallel to the fast axis (blue) and one parallel to the slow axis (orange). These components will oscillate as the electric field of the light oscillates, and this is shown in pale orange and pale blue for clarity. The component parallel to the fast axis experiences the lowest refractive index and travels fastest, and the component parallel to the slow axis experiences a higher refractive index and travels more slowly. As the waveplate is a half-wave plate, the slow component is delayed by a half-wavelength, such that the output electric field of this component (upper orange line) is 180° out of phase compared to the input field (lower orange line). The total effect is that the overall output polarisation (upper green arrow) now makes an angle of $-\theta$ to the fast axis. Therefore, **by rotating the waveplate by an angle θ, we rotate the polarisation of the incident light by 2θ**.

Another very common use of waveplates is the use of a quarter-wave plate to turn linearly-polarised light into **circularly polarised light**. Circularly polarised light has a rotating polarisation vector, which rotates around the propagation direction of the beam in a certain direction[11]. This arises in an analogous way to that described above, where delayed phase components cause the change when the waveplate is at certain angles. However the analogous diagram is substantially more complex! Good theoretical treatments are given in references [1, 2] but practically it can be achieved by placing the QWP with the fast axis at 45° to the incoming linear polarisation.

---

[11] This direction is called the 'handedness' of the polarisation, so you can have 'left-handed' or 'right-handed' circular polarisation. Unfortunately, the handedness depends on whether the light is going towards you or away from you—and no convention has been agreed!

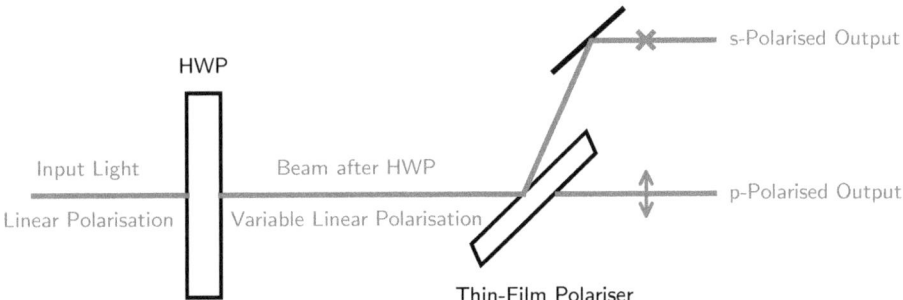

Figure 7.9. Schematic of a typical variable attenuator constructed using a half-wave plate and a polariser. Linearly-polarised input light is rotated to an arbitrary linear polarisation by the HWP, and the polariser then separates this into s-polarised and p-polarised components. Varying the angle of the HWP will vary the proportion of the input power sent into the s- and p-polarised output beams, allowing the power in a chosen beam to be varied.

Application: variable power control

One of the most common uses of a polariser in ultrafast laser labs in the author's experience is to provide variable power control (or *variable attenuation*) of a linearly-polarised laser beam. This is easily achieved by using a half-wave plate and a thin-film polariser together, as shown in figure 7.9. The incoming beam first passes through the waveplate, which can rotate the polarisation arbitrarily, and then is incident on a thin-film polariser. If the waveplate is rotated such that all the incident light is s-polarised, then all of the light will be reflected from the polariser, and vice versa. The end result is that two beams are produced (one from the reflection, and another from the transmission), and the ratio of the reflected:transmitted power is controlled by rotating the waveplate.

This setup for power control is relatively easy to build, but pre-built assemblies to perform this variable attenuation are also widely available from different manufacturers. The advantage of using thin-film polarisers here is that they can withstand high power, and therefore this setup can be used to attenuate high power beams. It is also ideal for ultrafast applications provided that a broadband TFP is used. A possible drawback of this setup is that if the polariser has a 99% extinction ratio, then 1% of poorly polarised light will come through even at the 'minimum' power. This 1% could be quite a lot of laser power if the input power is high enough, so stacking multiple attenuators together such that the output of one feeds into another is a good way to get very precise control of the output power. Waveplates can also be mounted in computer-controlled rotation mounts that allow very fine, repeatable tuning of the power. For example, you could build a two-stage attenuator using a manually adjustable 'coarse' stage initially, and then a computer-controlled 'fine' stage, to obtain very fine, very repeatable control over the output power.

7.5 Focussing optics

Generally the output of our lasers will be a beam of light which is well **collimated**. This means that it has a very low divergence angle, or that it doesn't spread out from its initial well-directed beam. However, there are times when we may not want our beam to be well collimated (for instance, if we want to change the beam waist—or focus the beam onto a sample), and in these cases we need an optical element to change the divergence angle—a **focussing optic**.

7.5.1 Lenses

The most common kind of focussing optic is well known in everyday life, a **lens**. A lens is a transmissive optic, and is curved such that as light passes through it, it undergoes refraction and the divergence changes. It is possible to make lenses that are both **converging** (focussing the beam down to a point), and **diverging** (expanding the beam). Lenses come in a variety of different shapes, but for now we will consider shapes where one side of the lens is flat, and the other side curved. These are called **plano-convex** or **plano-concave** lenses depending on whether or not the curved side bulges outwards away from the centre of the lens (convex), or caves back in towards the centre (concave)[12]. The main parameter that defines a lens is the **focal length**. The focal length, f, is defined as the distance from the lens that an incoming perfectly collimated beam will be focussed to—and is illustrated in figure 7.10. Lenses with short focal lengths are often referred to as 'strong' lenses, as they focus the light beam very hard and to a very small point (and vice versa for longer focal length lenses). A diverging lens has a negative focal length, defined as the location that an image *would be formed at* if the outgoing beam was extrapolated back behind the lens. This is sometimes called a *virtual focus* or *imaginary focus*. The **optical power** of a lens (measured in **dioptres**) is defined as the reciprocal of the focal length, and this is the number an optician prescribes you when giving a glasses prescription. Stronger lenses have higher optical powers. Lenses made of material of a higher refractive

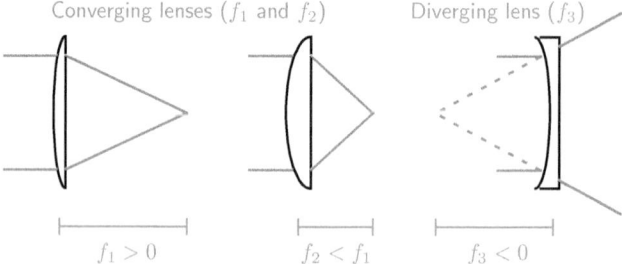

Figure 7.10. Illustration of three different focussing lenses. From left to right: a converging lens with positive focal length f_1; a stronger converging lens with positive focal length $f_2 < f_1$; a diverging lens with negative focal length f_3. In all cases collimated light enters the lens on the left and leaves on the right. Dashed lines on the diverging lens show the position of the virtual focus. Note that collimated light hits the curved side of the lens first—this is to minimise spherical aberration.

---

[12] Lenses with one flat side are most often used when the beams entering and leaving the focussing system are well collimated, and this is most often the case in (at least) the majority of experiments encountered by the author. Other lenses may be more suitable for your application—particularly if you are using the lenses to perform imaging rather than to just manipulate beam divergence.

index will be more powerful, and lenses that are more strongly curved will also be more powerful.

The exact focal length of a lens generally depends on the colour of the light being focussed, as refractive index depends on wavelength, so (for example) blue light will generally experience a higher refractive index than red light in most optical materials. This means that a lens with a certain focal length for blue light will actually have a longer focal length for red light. Put another way, the same lens will tend to focus blue light harder than red light. This focussing of different colours to different points is a form of **chromatic aberration**[13], and can have important consequences in spectroscopy when overlapping foci of beams of different colours as will be discussed in due course. However, it is not generally the case that the focal length of a lens will vary appreciably over the bandwidth of an ultrashort laser pulse, except in the most extreme circumstances.

However, lenses are transmissive and so do add dispersion to any optical assembly, which can be problematic for ultrashort pulses. A good upper estimate for the amount of GDD a lens will add is the same as the amount a flat piece of the same material as thick as the widest part of the lens would add. If the GDD added by a lens would be problematic for your application, then the solution is to use a **focussing mirror**. When working with lenses it is important to keep in mind that focussed high intensity laser beams are much more hazardous than non-focussed beams and will much more easily burn things (including your skin), so you should always place blocks after lenses until you are certain the beam is going where you want it to. Additionally, it is important to ensure that when passing a beam through a lens, that you hit the middle of the lens, and that you have the lens placed such that it is exactly orthogonal to the propagation direction of the incoming beam. Having a lens off-centre or at an angle can introduce severe aberrations that are difficult to correct for later.

Application: focussing laser beams
The simplest situation where lenses are used is in focussing laser beams onto experimental samples. This is often necessary as the beam from a laser output is often quite large, and so just shining this beam onto a sample will often not provide enough intensity at the sample to perform the measurement of interest. A simple and cheap way to increase the intensity is to make the beam waist at the sample smaller. A typical example might be taking the output of a Ti:Sa laser system with a beam waist of around 5 mm, and focussing it using a lens to a point with a beam waist of around 50 μm. Reducing the beam waist by a factor of 100 (as here) will result in the irradiated area being reduced by a factor of 100^2—increasing the intensity by a similar factor.

To do this in a lab, you need to have a way to calculate the focal length of the lens needed to produce a given beam waist in the focus. More details on the full theory of Gaussian beam focussing can be found in many standard texts [2–4], and online

---

[13] Chromatic aberration can be minimised (or eliminated) through use of an **achromat**—a compound lens made up of multiple lenses that compensate for the aberration.

calculators for many of the relevant parameters can be found in the Optics Toolbox [5]. However, if we assume that our lens is placed well inside the Rayleigh range of the beam[14], then the beam waist after focussing x_0' is given by equation (7.1).

$$x_0' = \frac{f\lambda}{\pi x_0},\tag{7.1}$$

where f is the focal length of the focussing lens, λ is the wavelength of the focussed light, and ω_0 is the beam waist before focussing. Because our lens is well inside the Rayleigh range of the laser, then x_0 is essentially the beam waist of beam as it hits the lens, i.e. the size of the beam on the lens. Equation (7.1) illustrates some key concepts in beam focussing.

1. Lenses with a longer focal length f will produce larger beam waists after focussing and vice versa. This is intuitive as we think of lenses with longer focal lengths as being 'weaker' lenses that don't focus as hard.
2. Longer wavelengths of light λ will focus to larger beam waists than shorter wavelengths. This is perhaps less intuitive but is consistent with our knowledge that blue light will refract more strongly than red light (hence the blue parts of a rainbow are closer to the inside than the red parts).
3. A smaller beam hitting the lens (smaller x_0) will focus to a *larger* beam waist ω_0'. This seems counter-intuitive, but can be understood by considering the construction of a typical lens. Close to the centre of the lens, the lens is less strongly curved than it is at the edges—therefore a beam hitting the edges of a lens (a larger beam) will be refracted more strongly than a beam hitting only the very centre (a smaller beam). The stronger refraction results in a more tightly focussed beam, and a smaller x_0'.

Some of these concepts can seem counter-intuitive at first, but will quickly become intuitive when you start working with lenses in the lab. We can now use equation (7.1) to calculate the focal length of light we would need to perform the aforementioned 5 mm to 50 µm focussing of an 800 nm beam. Plugging these values in results in a required focal length of around 1 m. Be aware that some online calculators for these parameters may define the beam waists in terms of diameters ($2x_0$) rather than radii.

Application: telescopes
Sometimes it is desirable to change the beam waist of a collimated beam. For example, the output beam from your laser may be large and difficult to work with, so making it smaller (whilst retaining collimation) is desirable. Alternatively, you may have a very small beam that you want to focus hard, and need to expand the beam so that it is larger when it passes through the focussing lens. In all of these cases, you want to change the beam waist whilst maintaining the collimation of the beam. This requires that you use a **telescope** of some description, either a **beam expander** (to

---

[14] This ensures the laser beam is not wildly divergent at the lens position, so the exact position of our lens along the propagation direction of the beam does not enormously change the size of the beam on the face of the lens.

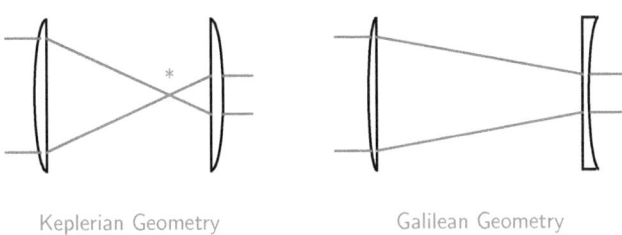

Figure 7.11. Schematics of a Keplerian (left) and Galilean (right) telescope, both set up to shrink an input laser beam (blue lines). In a Keplerian telescope the beam goes through a focus at the starred point, which can be problematic when using high intensity beams.

enlarge the beam) or a **beam reducer** (to shrink the beam). A telescope made using lenses is called a *refracting telescope*, and there are a few different designs available, of which the two most common are the **Galilean Telescope** and **Keplerian Telescope**.

Figure 7.11 shows an illustration of these two geometries of telescope. The Keplerian telescope (left) uses two converging lenses, whereas the Galilean (right) uses a converging and a diverging lens to achieve the same behaviour. The major disadvantage of the Keplerian design for this purpose is that the beam goes through a focus (indicated with a star on figure 7.11) between the two lenses. For a high intensity beam (and most of the time our ultrafast lasers produce high intensity beams), going through this focus can result in the intensity being high enough to ionise the surrounding air. This will (at least) ruin the beam quality after the focus, and if the intensity is high enough then non-linear frequency conversion can occur in the air which will make your beam look very colourful after the focus! For this reason, we tend to want to work with **Galilean** telescopes in ultrafast optics, and this is what will be considered further in this section.

A Galilean telescope that is set up to work as a beam reducer (shrinking a beam) will have a converging lens at the input and a diverging lens at the output (as shown in figure 7.11). Conversely, the same lenses could be used to produce a beam expander if the diverging lens was placed at the input and the converging lens at the output. To illustrate how to choose the specific lenses, we will imagine we want to create a beam expander that will increase the diameter of our input beam by a factor of two.

Figure 7.12 shows a typical design for such a telescope. Increasing the size of the beam by a factor of two requires that $d_2 = 2d_1$ and therefore $d_2/d_1 = 2$. The **magnification factor** of the telescope (M) is given by:

$$M = -\frac{f_2}{f_1} = \frac{d_2}{d_1}, \tag{7.2}$$

where f_1 and f_2 are the focal lengths of the input and output lenses, respectively, and d_1 and d_2 are the diameters of the input and output beams, respectively[15]. The minus sign in equation (7.2) is present because one of the focal lengths will be negative (that

---

[15] One may say that this should be defined in terms of beam waists and not diameters. However it doesn't matter in this case as we are only interested in the ratio, and it is a good lesson in dealing with people using non-standard units.

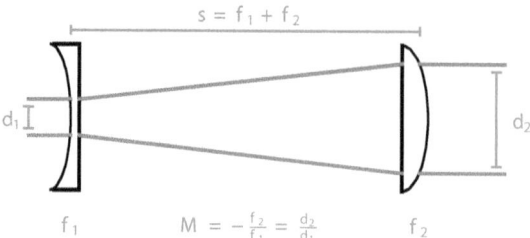

$$s = f_1 + f_2$$

$$M = -\frac{f_2}{f_1} = \frac{d_2}{d_1}$$

Figure 7.12. Design of a typical Galilean beam expander. Edges of a laser beam are shown in blue. A diverging lens with focal length f_1 is placed at the input and a converging lens with focal length f_2 re-collimates the output. The laser enters with a diameter d_1 and exits with a diameter d_2. The magnification factor M of the telescope is given by $M = -\frac{f_2}{f_1} = \frac{d_2}{d_1}$. The lenses should be placed at a distance $s = f_1 + f_2$ to ensure that the output is collimated.

Table 7.3. Lens combinations for beam expanding/reducing telescopes.

f_1 (mm)	f_2 (mm)	$s = f_1 + f_2$ (mm)	$M = -\frac{f_2}{f_1}$
−50	100	50	2
−100	300	200	3
100	−50	50	0.5
200	−50	150	0.25

of the diverging lens), but a negative magnification would be unphysical. As we know our ratio d_2/d_1, then we know that the magnification required for our factor-of-two expansion is simply 2. We can then pick any combination of lenses f_2 and f_1 to achieve this, with the caveat that the sum of the focal lengths must be a positive number (otherwise the distance between the two lenses would need to be negative). For example, this could be achieved by having $f_1 = -50$ mm and $f_2 = 100$ mm, such that $s = f_1 + f_2 = -50 + 100 = 50$ mm. Table 7.3 gives some further examples of combinations of lenses that can be used to give a variety of beam expanding telescopes (where $M > 1$) and beam reduction telescopes (where $M < 1$).

Constructing a telescope can be slightly fiddly, but mounting one lens on an inexpensive manual translation stage to allow the distance s to be adjusted so that the collimation can be fine-tuned is a good idea. Putting collars on the lens posts so that the angle of the lens relative to the incoming beam can be easily adjusted is also useful, as it is very important that the beam coming into the telescope hits the lenses centrally and square (so that the face of the lens is orthogonal to the propagation direction of the beam). When assembling the telescope, putting in the converging lens first is generally easiest, as you will be able to keep track of where the beam is more easily than if you put the diverging lens in first, but take care to block the focussing beam before it can damage anything. Placing the converging lens, and then adjusting the position and angle so that it is well placed, then placing the diverging lens, is the author's preferred approach. Ideally in this case the beam comes out collimated and without the pointing of the beam through the remainder of the beamline being substantially changed.

7.5.2 Mirrors

An alternative option for focussing laser beams is to use a **focussing mirror**. This is essentially a concave mirror that causes light reflected from it to be focussed down to a point. This is illustrated in figure 7.13. There is an immediate clear advantage of using a mirror to focus ultrashort pulses, which is that the mirror is not transmissive, so it will not add GDD to the pulse in the way that a typical lens would[16]. In addition, the focal length of the mirror does not depend on the colour of the incident light in the way that the focal length of a lens does. To understand this, consider that the root cause of lenses to focus different colours of light to different focal points (the *chromatic aberration* of the lens) is that the refractive index of the lens material varies with frequency. This means that different colours will undergo differing degrees of refraction through the lens and therefore be focussed to slightly different points.

Clearly, a focussing mirror will not suffer from this problem. As the mirror is not transmissive, then we don't care about the refractive index of it—and any colour of light incident on the mirror will be focussed to the same focal point. Practically, the lack of chromatic aberration means that focussing mirrors are ideal for use in places where focussing of many different colours of light is desired (for example, if you scan your laser output wavelength across a broad range, or need to focus many different colours to the same point on an experimental target). It is also possible to make **reflecting telescopes** using a curved mirrors. These are much less common than refracting telescopes as they are more fiddly to align and build, but can be useful if you cannot compensate for the accumulated GDD that a refracting telescope will introduce. Another advantage of focussing mirrors is that they can be used to focus high intensity beams that would cause self-focussing (or other non-linear effects) if focussed using a transmissive optic like a lens. Focussing mirrors are commonly specified using the **radius of curvature**, R, of the mirror rather than the focal length, f, but these are easily related with the formula $f = -R/2$.

It may sound as though using a focussing mirror is somewhat of a no-brainer when it comes to focussing your ultrashort pulses! No accumulated GDD and no chromatic aberration can be very useful; however, a lot of the time you will still

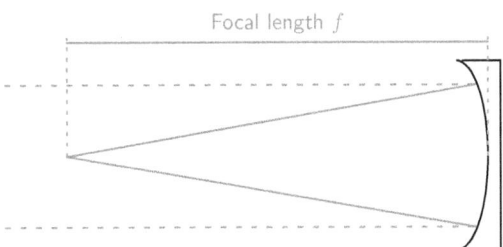

Focal length f

Figure 7.13. Illustration of how a focussing mirror works. The incoming beam is shown in dashed blue, and the focussed beam in solid blue. The focal length is defined from the centre of the concave mirror.

---

[16] This is with the exception of any GDD accumulated during the reflection—but focussing mirrors tend to be metallic and so incur minimal GDD anyway.

encounter people using lenses for focussing unless they absolutely have to use mirrors. The reason for this is two-fold. Firstly, focussing mirrors can be very fiddly to align properly. Having a single optic both reflect the beam and focus it can be difficult enough to align that many people would rather opt to decouple these effects and use a mirror followed by a lens, and then compensate for any extra GDD if necessary. Secondly, focussing mirrors tend to be made of metal, and have the same downsides as all metallic mirrors. That is, they are more easily damaged and less reflective than dielectric mirrors. For these reasons, use of lenses is still widespread, and focussing mirrors are reserved only for those applications that truly demand their use.

7.6 Gratings and prisms

Now we turn to consider optics that can spatially disperse different colours of light, **diffraction gratings**, and **prisms**. The dispersive action of a prism will be familiar to many from Isaac Newton's famous experiments, or the cover of Pink Floyd's timeless album 'Dark Side of the Moon', and the rainbow effect that can be seen on the back of a CD/DVD is because the lines etched into the disc that store the data function as a diffraction grating and disperse the light. It is important to note that in this case when we say 'dispersive', we are talking about **spatial dispersion**, the spreading out of different colours *in space*; as opposed to **chromatic (temporal) dispersion**, which is the spreading out of different colours *in time* that we have already discussed at length (and referred to simply as 'dispersion') in this book. Going forward, I will be explicit whenever I refer to dispersion in a potentially ambiguous situation, but assume that an unqualified 'dispersion' refers to temporal dispersion, as it has for the book so far.

Being able to spatially disperse the colours within our laser pulse is a very useful thing in that it allows us to essentially decompose our broadband laser pulse into its constituent frequencies. This is known as *accessing the Fourier plane* by aficionados[17], as we have physically moved from working in the time domain to the frequency domain after the spatial dispersion, just like we would if we performed a Fourier transform. Accessing frequency space like this allows us to do a plethora of useful things to our laser light such as pulse compression and stretching, as well as being very useful diagnostically.

7.6.1 Gratings

A **diffraction grating** (referred to as simply a 'grating' most of the time), is an optical substrate which has a large number of very fine lines physically etched into the surface of it. Light can then be bounced off of the etched surface (a 'reflection grating'), or passed through the back of the substrate and then through the etching (a 'transmission grating'). The spatial dispersion that is caused when light bounces off or passes through a grating depends on the density of the etched lines, and so

---

[17] Strictly this requires use of an extra focussing element, but we are still accessing frequency space by dispersing the colours.

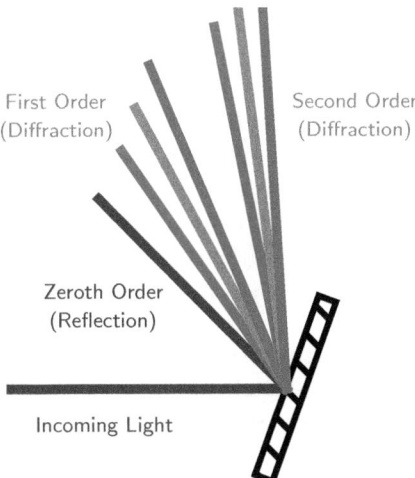

Figure 7.14. Illustration of reflection of a laser beam from a typical diffraction grating. The diffraction gratings produces various diffraction orders. The first and second orders spatially disperse the frequency components of the light (the light is diffracted), whereas the zeroth order is simply reflection from the grating surface.

manufacturers specify the number of lines per unit length, generally as lines per millimetre. A higher density of lines causes more severe spatial dispersion.

Figure 7.14 shows diffraction from a typical reflective diffraction grating. The reflection consists of multiple diffraction orders of which three are shown (this may be familiar in the context of X-ray diffraction from crystal structures). The first and second order diffractions spatially disperse the light, while the zeroth order diffraction is just a reflection from the grating surface. Having multiple diffraction orders can create a lot of stray beams and power loss, so it is often desirable to suppress unwanted orders. This can be done by engineering the shape of the etched lines in the grating, which is a process known as **blazing**. A **blazed grating** will concentrate light of a specific wavelength into a certain diffraction order, enhancing the efficiency in this region. For example, a grating may have a 400 nm blaze, which means that light incident on the grating around the 400 nm region will be concentrated into a specific diffraction order—and so the grating is most efficient when used at this wavelength.

Gratings are commonly found inside laser systems for pulse compression. They are preferable to prisms for this purpose because they can cope more easily with the high intensity beams inside a typical output compressor, and also add much more GDD. However, gratings are generally more expensive than prisms, do not function over as wide a range of wavelengths, and can cause significant power loss.

7.6.2 Prisms

A **prism** is a wedge-shaped piece of glass (shaped like a trigonal prism), which spatially disperses light. The spatial dispersion of a prism is caused (again) by the refractive index of the prism material being dependent on the wavelength of incident light. Different wavelengths of light experience different refractive indices and so undergo different amounts of refraction in the prism, which leads to spatial dispersion. Accordingly, the material that the prism is made from affects the degree of spatial dispersion (higher refractive indices disperse more strongly).

A cursory look at any manufacturer's catalogue will show that a lot of prisms are intended for beam steering or imaging (binoculars and spotting telescopes tend to use roof-prisms in their construction) rather than simply spatial dispersion. Prisms are generally inappropriate for beam steering with ultrashort laser pulses as the GDD added by the prism material will broaden the pulse significantly. When working with ultrafast lasers, prisms are generally only used for their dispersive properties—to make compressors/stretchers, to tune the dispersion in a laser oscillator cavity, or to spatially disperse a beam for diagnostic reasons.

Application: beam diagnostics
Using a cheap prism can be an excellent way to spatially disperse the frequencies in the pulse when doing laser diagnostics. A simple illustrative example of how this can work is when attempting to look for a non-linear interaction. For example, you may be using a non-linear crystal to generate the second harmonic of your laser output. It can be difficult to see the second harmonic beam as it could be relatively weak and will overlap in space with the fundamental. Directing both beams into a prism and catching the spatially dispersed beams on a block will separate the colours, allowing you to clearly see the second harmonic (and optimise your crystal parameters to maximise how much you have).

7.7 Windows and filters

The last types of optical elements we will discuss are windows and filters. Windows are transmissive elements that allow light to pass through them. They are a necessity if you are bringing your beams into a vacuum chamber, or passing them into some other piece of experimental apparatus. A filter is essentially a window that only allows a select range of frequencies to pass through it—filtering out unwanted frequencies. The classic use of a filter is to isolate a small signal of one colour against a much more intense background signal of a different colour. A special kind of filter is the **neutral density filter**, which will be discussed shortly.

Both filters and windows are transmissive, and so will add some dispersion to an ultrashort pulse. This is generally unavoidable, as you cannot (for example) remove the window from your vacuum chamber to make the pulses shorter! As such, pre-calculating the added GDD from these elements is a good idea, and then you can more easily pre-compensate for it if necessary.

7.7.1 Windows

Windows are perhaps the most simple optical element—a flat piece of glass that allows light to pass through it. Clearly, the important considerations are of the window material and the coating. The window material should be chosen so that it has good transmission properties at the desired wavelength, and does not incur too much GDD. Windows can also be AR coated, which is generally a good idea if your application allows it.

The only real complications in using windows arise because windows are often a structurally integral part of experimental apparatus, especially in high vacuum experiments. In this kind of experiment, the window forms the interface between atmospheric pressure and an ultra-high vacuum. In this case the window needs to be thick enough to not be easily broken, and also needs to be mountable in a vacuum flange. A 3 mm window is normally strong enough without causing terrible pulse broadening. Various options for mounting windows to vacuum chambers are available from manufacturers. An additional consideration is that if you are focussing a beam into a chamber, take care to ensure that the beam is not so small on the window that it will drill a hole in it over time—this will rather tend to spoil the vacuum!

7.7.2 Filters

A **filter** is ultimately just a window with a coating that will only allow certain colours to pass through it. There are a wide variety of different kinds of filter: long- and short-pass filters which will transmit all wavelengths longer or shorter than a certain cutoff wavelength; band-pass filters which will transmit all wavelengths in a specified range; wedge filters which block all wavelengths except for those in a narrow range (the 'wedge'); and notch filters, which transmit all wavelengths except for those in a narrow range (the 'notch'). Filters can work by either reflecting or absorbing the non-transmitted light. Reflective filters are less easily burned (they don't have to absorb large amounts of light), but also create more stray beams that need to be properly blocked.

The quality of a filter is determined by how much of the undesired light is able to pass, and also by the 'sharpness' around the cutoff wavelength. A higher-end filter will be able to cut much more sharply, and will block more of the undesired light. Filters are also made of glass, so the usual considerations about GDD also apply.

Application: isolating signal from background beams
Many kinds of non-linear spectroscopy require beams to be overlapped onto a target, and their combined interaction inside the target produces a beam of a different colour as the desired experimental signal. This beam is generally a much lower intensity than the other two beams, so the other beams need to be blocked for the signal beam to be recorded. A filter is ideal for this job, as illustrated in figure 7.15.

7.7.3 Neutral density filters

A special kind of filter that deserves a separate section is the **neutral density filter**, or **ND filter**. ND filters act to attenuate all colours in their (very broad) bandwidth equally (hence the 'neutral'), so are ideal for use when simple and fast reduction of laser power is needed, such as for alignment or diagnostics. They too come in either

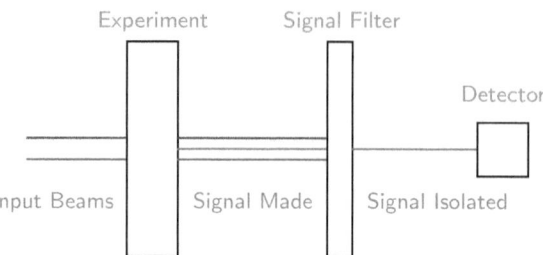

Figure 7.15. Schematic of how a filter can be used to isolate a desired signal beam (green) from collinear unwanted input beams (red and blue).

reflective or absorptive variants, and can be made of a variety of substrates to suit different applications. ND filters are fairly easily burned, but are also fairly cheap and come in kits containing multiple filters. The degree of attenuation is marked on the filter as the **optical density**, and this is generally a number between around 0.1 and 10. The power of a beam passing through an ND filter with a given optical density (OD) is reduced by 10^{OD} after transmission. For example, an ND filter marked with an OD of 2 will reduce the power by factor of 100. Multiple filters can be stacked to provide greater attenuation, and in this case the largest (highest OD) filter should be the one that the beam hits first, to minimise the possibility of damage.

A useful type of ND filter is the **variable ND filter**, or **ND wheel**. These are circular filters where the optical density increases around the circumference of the wheel, and can be used to provide a straightforward way to variably attenuate beams. This is especially useful in the UV region, where the waveplate-polariser variable attenuator previously mentioned is difficult to make and would add a lot of GDD. It is also useful if using a widely tunable source, where a waveplate-polariser attenuator would not function over a wide enough bandwidth to be useful.

7.8 Optomechanics

The final thing to discuss in this chapter is **optomechanics**, which is a catch-all term for all of the things that hold and manipulate the optical elements on a laser table. A lot of manufacturers of optics also manufacture optomechanics, and there is a dizzying array of different things available to allow construction of even the most complex beamlines. Experienced campaigners will have their own opinions as to who makes the best optomechanics, so asking around your lab/department is well worth the time spent. Many sell pre-assembled mounts for things like periscopes and delay stages, which can save a lot of time and hassle making and aligning these assemblies from scratch (albeit at an increased cost). Exhaustive discussion of all the different types of optomechanics will not be given here, but a short overview is given together with some potential pitfalls to avoid when buying.

7.8.1 Anatomy of a mounted optic

Figure 7.16 shows a CAD drawing of a mounted optic. In this case, it is a mirror mounted in a kinematic mount, attached to a post and post holder, which are

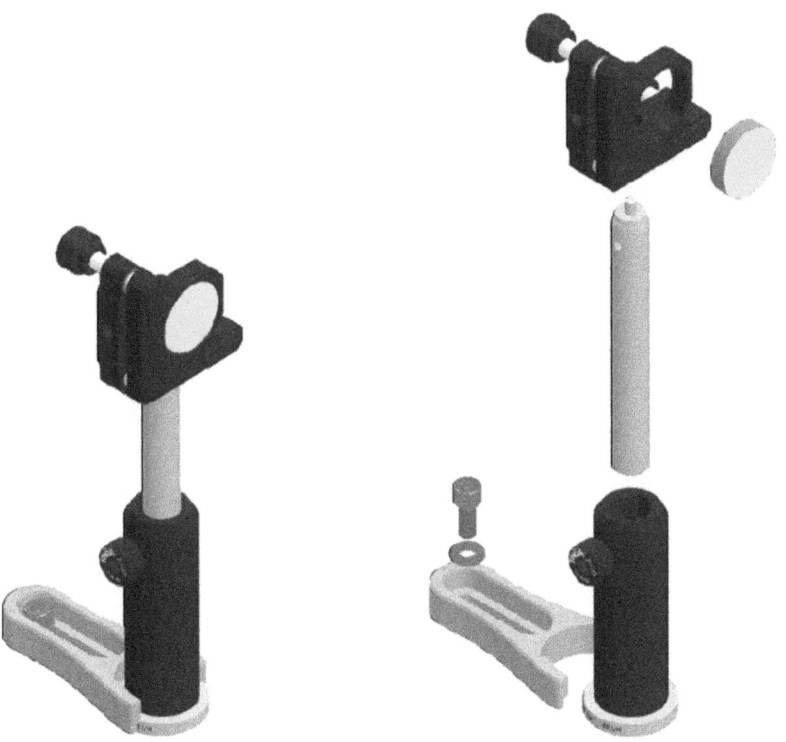

Figure 7.16. CAD drawing of a typical mounted mirror shown both complete (left) and exploded into constituent parts (right).

clamped to a laser table. We will discuss this from the top down to illustrate the purpose of mounting optics in this way.

Initially, the mirror (here shown with a gold face) is mounted into the mirror-shaped recess in the kinematic mount. The mirror is secured in place using a small set screw (normally an M2 screw[18]), which is tightened with a hex key. These set screws tend to have a small piece of rubber or felt on the end of them to avoid scratching the optic. Care should be taken to not over-tighten these set screws, as doing so can cause very thin optics to warp—which can cause the beam to inadvertently focus or become elliptical. The entire kinematic mount is attached to an optical post—in this case a 1/2 inch stainless steel post—with either an M4 set screw or M4 socket head bolt. The post has a hole in the body of it that a hex key can be inserted into to aid in tightening the mount to the post.

The post is the inserted into a post holder. The post holder is anodised black to minimise reflectivity and the post is tightened inside the holder using a thumbscrew. The post can be slid up and down within the post holder to allow the height of the mirror to be adjusted, and can also be rotated within the holder for coarse

---

[18] The ISO metric screw thread system is most commonly used. The number refers to the outer diameter of the thread in millimetres. An M2 screw has a 2 millimetre diameter thread.

adjustment of the mirror angle (fine adjustment is taken on the thumbscrews on the kinematic mount itself). The post holder can be attached to the table in a variety of ways, but the most convenient way is through using a pedestal mount and clamping fork. The pedestal mount has an M6 stud on it which screws into the base of the post holder. The optic is then placed in the desired position on the laser table, and a clamping fork is placed over the pedestal. The fork is then bolted to the table using an M6 socket head bolt. This holds the entire mounted optic securely in place, but also makes moving the optic by undoing the bolt in the clamping fork relatively straightforward. Various lengths of fork are available to suit different requirements.

There are other ways to mount optics than this, but they all follow roughly the same principles—a holder is securely bolted to the table, and the optic is attached to a post which is placed in the holder that allows the height and angle to be easily adjusted. Some optical mounts do away with the separate post holder and simply clamp a thick optical post straight to the table. This is more stable, but does away with any easy height or angle adjustment—it is commonly found within laser cavities where the stability is necessary.

7.8.2 Tips for buying optomechanics

When buying optomechanics you should first consider what kind of stability you need for your specific setup. Most manufacturers make a range of different 'tiers' of optomechanics, often with a standard, cheaper tier, and a higher-end expensive tier. Optomechanics from the higher-end tiers are made with greater precision and are designed to give better long-term stability; and mounts that feature kinematic adjustment (i.e. the ability to move the optics) will allow finer and more repeatable control. The downside is a vastly increased cost, so if these are not truly needed, then using cheaper optomechanics can be better. It is possible to mix-and-match, so that you use a high-end ultra-stable mount for overlapping your beams (allowing you finer control of the overlap position), but cheaper mounts elsewhere. Specifications and benchmarks for the stability of things like translation stages are available from manufacturers, so you can see if they will work for your application.

Another thing to consider is ease of building. I would generally advocate that penny-pinching on cheap things that are more difficult to use can be a false economy—the cost of the time you can spend trying to build things from scratch can often outweigh the cost of buying something pre-assembled in the first place. This is especially true in optomechanics, and so spending slightly more money on (for example) a mount for delay stage mirrors that is cut at exactly 90° is generally a wiser investment that possibly having to spend many hours ensuring that your two individual mirror mounts are precisely at 90°. Of course, sometimes budgetary constraints require you to be more creative, but do not feel that spending extra money to make life easier for yourself is somehow a cop out—we are non-specialists trying to use lasers to answer our research questions, not laser physicists!

A final thing to be aware of are the standards for optic sizes. Optics of a 1 inch diameter are commonly used, but many other sizes are also used, so ensuring that you're buying the correct mounts is important. Equally, if mounts attach to posts and holders using screws, then ensuring that you have the right screw threading is critically important. Whilst most of us in the world use ISO standard metric 'M' designated screw threads, people in the USA commonly use the ANSI standard UNC and UNF imperial screw threads—and the threads are generally not compatible with each other. Manufacturers make things for both markets, and so will normally offer both imperial/metric options on most items. Bitter experience tells me that taking the time to check you're ordering the right things can save a lot of future headaches!

References

[1] Goldstein D H 2017 *Polarized Light* 3rd edn (Boca Raton, FL: CRC Press)
[2] Pedrotti F L, Pedrotti L M and Pedrotti L S 2017 *Introduction to Optics* 3rd edn (Cambridge: Cambridge University Press)
[3] Paschotta R P 2014 *The RP Photonics Encyclopedia* www.rp-photonics.com/encyclopedia. html (Accessed: 22-12-2020)
[4] Paschotta R P 2008 *Encyclopedia of Laser Physics and Technology* 1st edn (Berlin: Wiley-VCH)
[5] Light Conversion *Optics Toolbox* http://toolbox.lightcon.com (Accessed: 23-12-2020)

Chapter 8

Building a beamline

This chapter focusses on a discussion of the basic principles of *building* an experimental **beamline**, using the optical elements described in the previous chapter. The term **beamline** refers to the arrangement of optics that allows you to control and direct the output of your laser system into your experiment. A beamline could be as simple as two mirrors directing the laser output into a sample cell, or could be very complex with multiple overlapping beams and lots of additional optical equipment to control various properties of the laser light. In all cases, there are some fundamentals that form the bedrock of effective optical design and construction, which will be codified here for reference.

Construction of beamlines is often a relatively infrequent part of everyday lab work—once a beamline is constructed and working, it is often left in place and used rather than continually rebuilt. As such, many students and researchers may work for several years without actually having to do substantial optical building work, and then often find themselves thrown in at the deep end when expected to know how to construct the thing they have been working with for many years! The aim of this chapter is to codify some methods of best practice into an accessible resource, so that somebody experienced with working with optics, but not necessarily experienced with beamline construction, can avoid some of the more common pitfalls.

8.1 Safety!

Before embarking on a discussion of how to work with optics, it is prudent to make a point about safely working with lasers and laser light. Whilst lasers are an incredibly useful and ubiquitous tool in experimental physical sciences, they are also **very dangerous if not used safely**. There is a worldwide system where lasers are rated from 'Class 1' to 'Class 4' depending on how powerful (and dangerous) they are. All lasers discussed in this work are Class 4 lasers, and the majority of lasers used in scientific research are either Class 3 or Class 4. These lasers have the potential to cause serious eye damage, if not blindness, if used incorrectly. A typical Ti:Sa laser beam is many billion times brighter than the Sun, and even a small reflection could cause serious eye damage or blindness.

All research environments should provide laser safety training, and have systems in place to minimise the chance of dangerous accidental exposure such as interlocks and shielding. There will also normally be a designated laser safety officer who will be able to advise on safe practice, and will probably have to sign off any substantial new optical construction as safe for use. Basic principles such as always wearing safety glasses/goggles, always aligning optics with as low a power as possible, and always taking care to block any stray beams, should be standard practice. The author of this book sustained a retinal burn in his left eye from an unpredicted reflection from a polariser, despite taking all the usual safety precautions, and was lucky to escape with only very minor lasting damage that doesn't impact everyday life. The overall message is that **laser safety should be taken very seriously**.

8.2 Planning

The first thing to do when building a new beamline is to make a plan of what you need to build. Thorough planning will help to foresee and avoid potential problems, so taking the time to make a detailed plan is worth it in the long run. There are three key questions to answer when starting to design a beamline.

1. What light do I need at my experimental target, and how different is this light from the direct output of the laser system?
2. How much space do I have to build in?
3. What budgetary constraints are there, if any?

We will consider each of these questions in turn, and then show an example case study in the final chapter to illustrate the process.

8.2.1 What light do I need?

This is the first question to ask yourself when building a beamline. As discussed in preceding chapters, there are a variety of different properties of the light that we can manipulate. The answer to this question will depend hugely on the exact experiment being undertaken, but the key idea is to determine how different the light needed at the experimental target is from the direct output of the laser system, as this will determine how much manipulation of the output is needed (if any). Clearly, the easiest situation would be one where the output of the laser can be directly steered into the experimental target using a couple of mirrors, but it is rarely this straightforward in practice! The parameters that need to be determined are as follows:

- **Beams**: How many beams do you need to overlap at your target?
- **Position**: Where is the target? Do you need to come in at a specific angle?
- **Colour**: What colour(s) of light need to be at the experimental target? Do you need to do any external frequency conversion of the laser output?
- **Size**: What size of beam do you need at the target? Do you need to focus the beam down, or expand it?

- **Pulse Duration**; How short do your pulses need to be at the target? Do you need to do any external compression or stretching?
- **Intensity**: What intensity do you need at the target? With a fixed beam size and pulse duration, this will determine how much power you need in the laser beam.
- **Polarisation**: What polarisation do you need at the target? If you need multiple beams, do they need variable polarisation?

Taking these points in turn, clearly requiring more beams to be overlapped at the target increases the complexity, as the beams will need to be overlapped both spatially (so that each beam irradiates the same area), and temporally (so that the pulses in each beam arrive at the same time, or with a known delay). Strategies for this are outlined in subsection 8.3.2. Where the beams need to be is also a critical consideration, as in some cases it may be that you have a well-defined immovable target that you need to hit (such as the window into a large vacuum system), or other times it may be more flexible, and you may be able to move the sample around as desired. If the height of the experimental target is substantially different to the height of the laser output, then you may need to install periscopes to either raise or lower the beam height as needed.

Requiring a different colour of light at the target than is produced directly by the laser system requires the use of external frequency conversion. This could be as straightforward as some frequency doubling using a single non-linear crystal (for example, if your laser produces 800 nm pulses, and you need 400 nm pulses at your target), or may require the use of an external OPA to allow the colour of the light to be more widely tunable. This is commonly the case in UV photochemistry, where photochemical pumping of a target across a relatively wide band of the UV spectrum is desirable. The conversion efficiency of these frequency conversion systems varies, but at wavelengths far from the pumping wavelength it is often very low. For example, pumping an OPA with around 3000 mW of IR light may produce only 30 mW of UV light. This is something to consider if you are buying a whole new laser system for your setup, as you may need to use some portion of the laser output to pump an OPA, whilst retaining some portion for use in other areas of the beamline. It could be an expensive mistake if you have insufficient output power from the system!

In a similar vein, requiring a different pulse duration than that produced by the laser output (or compensating for dispersion incurred as the pulse travels towards the experimental target) can create some challenges. Often, the pulses produced by the laser are sufficiently short, and dispersion through the beamline is minimal, so that no additional compression or stretching is necessary. However, especially with UV pulses, the dispersion incurred through the beamline can be severe enough to require re-compression of the pulse before it is used in the experiment. Alternatively, you may wish to stretch the pulses (for example, if you want to reduce intensity but not change the power or spot size), and this will require construction of a stretcher. In both cases, building a compressor/stretcher can be a considerable expense—so it is worth taking the time to plan and calculate if one is needed so it can be factored into your budget.

The other parameters are more easily dealt with in the beamline construction. The required beam size at the target will depend on the nature of your experimental target. You may have a very dense sample, allowing you to focus the beam hard into the target and reduce the power in the beam. Conversely, the sample may be more diffuse (or you may want to use larger beam sizes for ease of overlapping multiple beams), so you want to irradiate a larger area of the sample to 'talk to' more of the sample with the light—but this may require more power in the beam. In all cases, the size of the beam at the target can be easily manipulated using lenses and telescopes. Similarly, the power of the laser beam can be easily manipulated using a variable attenuator such as a HWP-TFP combination or rotatable ND filter. Polarisation too can be varied using a relatively inexpensive waveplate, or combination of wave-plates. There are more complex situations where very fine control of polarisation is required over a broad frequency range, but these are relatively rare.

Having established the properties of the light that we require, we can then start to plan out the arrangement of optics on the laser table.

8.2.2 General principles

Planning a beamline is the first step of building a beamline, and important decisions are made in the planning stage that will affect the future operation. We will give an example of the planning of a beamline for a simple experiment from start to finish in chapter 9, and so here will aim to codify some principles of 'best practice' to keep in mind when planning.

Power control
The direct output of our laser system may well produce vastly more power than we need in the experiment, so it needs to be attenuated somewhat. It is best to try to attenuate this power **earlier**, rather than later. The reasons for this are two-fold:

- You will more often be working with the beamline nearer to your experiment than the output, so attenuating the beam closer to the laser output means that high power beams are not flying around the space where you will often be working, unless absolutely necessary. It is much safer to work with the beams attenuated, and excess power safely dumped, in an area of the table you will not often be standing at.
- Higher power beams have more potential to damage optics, or cause unwanted non-linear effects like self-focussing. Dumping the power earlier minimises the possibility of this.

The safest way to dump excess power is to use a purpose built **beam dump**, which is rated to handle the amount of power you need to get rid of. Very high power beams may require the use of a water-cooled beam dump, but this is unlikely in a typical spectroscopy lab.

Beam height and stability
In general, it is safest to have all the laser beams propagating in a single plane that is parallel to the surface of the optical table, and is at around waist height.

This minimises the chance of stray reflections being reflected upwards towards our eyes. In an ideal world, then, the output port of the laser system and the height of our experimental sample will be the same, and all the beams will propagate in a plane at this height. In reality, this is not normally the case, and so adjusting the height of the beam is necessary. Modifying the height of the beam can be accomplished either using a **periscope**, or by simply shooting the beam out-of-plane upwards or downwards, to another mirror in a different plane. If the height difference is relatively small, then the second option can work fine, but for large changes in height the use of a periscope is preferable. It is imperative that periscopes are properly shielded, as beams coming vertically up from the table have the potential to cause serious injury.

In addition, it is generally preferable that beams run closer to the table surface, as this minimises the length of the optical posts holding optics, which increases stability[1]. If the beam has to propagate a long distance between the laser output and experimental target, then it can be especially advantageous to periscope it down to a low height, and then bring it back up nearer the region of interest. As in some cases any small instability a long way 'upstream' can cause a larger instability further 'downstream', so maintaining stability over long propagation distances becomes critical. More expensive optical mounts are also more stable, but come at an increased cost. Shielding the laser and beam path from air currents in the room can also help stability enormously, and having the beam completely enclosed is much safer too.

Beam size
The size of the laser beam dictates what diameter the optics in the beamline need to have, and this can be an important consideration as large optics can be considerably more expensive than small optics. Typical Ti:Sa laser systems can produce relatively large beams, so the direct output may be too large to comfortably fit on 1 inch optics, and require the use of large 2 inch mirrors or beamsplitters. In this case, it can be advantageous to use a telescope to reduce the beam diameter such that it can readily fit onto a 1 inch optic, as constructing an entire beamline using 2 inch optics could quickly get prohibitively expensive. A telescope would also help to control any divergence in the laser output if propagation over a long distance is desired. However, an important consideration regarding beam size is that a larger beam can be appropriate for very high power beams, as the larger beam diameter spreads the power over a larger area, minimising the chances of optical damage.

Ease of alignment
A final consideration when making an initial plan for an optical beamline is to consider how easy it will be to align on a day-to-day basis. Laser systems drift, and the pointing stability of the laser may not stay constant over time. This means that even if you perfectly aligned all of the optics in the beamline once, they will probably not stay perfectly aligned and will require periodic readjustment. The beamline can

---

[1] A longer post has a longer moment arm between the optic and table.

be built to maximise the stability by running beams nearer to the table, and using heavier duty (more expensive) optical mounts; but it is also important to build the beamline such that getting it back into alignment *after* it has drifted is as pain-free as possible. There are a few principles which help achieve this aim:

- The optical table has a grid of M6 tapped holes on it, and making sure that (as far as possible) beams run along the lines of the bolt-holes, and that beams run orthogonal to each other, will make it easier to see when the beam has drifted out of alignment, and give you a rough target to aim for to get it back into alignment.
- Using alignment targets or irises in critical areas can help you get and maintain precise alignment. Irises are especially useful as you can periodically close the iris to check that it is closing centrally around the incoming beam, providing a very fast way to gauge if things have drifted. Critical areas are places where the alignment of the beam path is paramount to the correct functioning of the experiment. Typical examples of 'critical areas' would be where the beam enters an experimental target, OPA, or delay stage. Screwing irises directly into the tapped holes on the table ensures that beams are aligned to the bolt-holes.
- Making sure that the beam is (as far as possible) centred on every mirror, such that you can more easily use the full range of the adjustment screws to fine tune the alignment.
- Having a beam which is substantially smaller than the diameter of the optics (at least half the diameter), so that the beam is able to be moved around on the surface of a mirror to a greater extent before the mirror needs to be moved.

Some of these points will become clearer once we discuss how to actually align optics in the next section.

8.2.3 Sketching a beamline

Bearing the above principles in mind, planning out the spatial arrangement of the optics in a beamline normally starts with a simple sketch, which is an ideal way to initially get a feel for how you will fit all of your optics in, and get an idea of the rough sort of size your setup will need. Often, doing a sketch like this, using reasonably accurate dimensions is entirely adequate for making sure you have sufficient space, but sometimes (in more constrained areas) being more accurate can be beneficial. Most manufacturers of optomechanics will provide dimensions of their items, and will often also provide CAD files that can be loaded into a 3D drawing program. These programs may seem like overkill for this kind of optical design, but if time permits then this is the best way to get a realistic feel for how your planned beamline will look on the real optical table. The author has used Autodesk Inventor to design an entire lab space, starting with a vacuum instrument and then including an optical table and beamline with all associated optics, as shown in figure 8.1. While this took a long time, it meant that construction of the instrument and

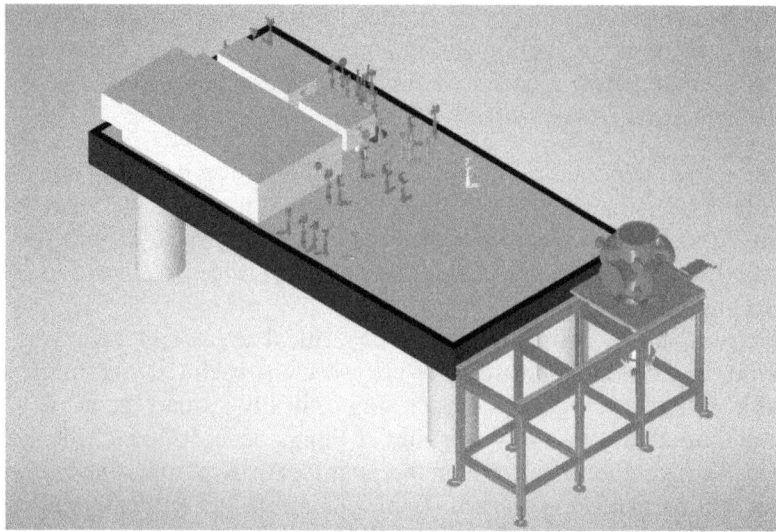

Figure 8.1. Initial CAD drawing for a lab space, including a multi-user beamline, designed by the author. Colours of the various optomechanics distinguish which part of the beamline they are intended to be used for.

beamline went a lot more smoothly than it would have had it not been planned so thoroughly. If time permits, this can be beneficial—but often a simple 2D sketch will suffice for most purposes. It is also important to remember that plans are made to be deviated from, and the final beamline in figure 8.1 ended up looking slightly different than the beamline as designed here. However, having a thoroughly planned place to start makes deviation easier, and minimises the chance of you encountering a totally insurmountable obstacle when building.

8.2.4 Budgetary constraints

Often, you will be working to some kind of a budget (depending on what kind of research environment you are in!). Optics and optomechanics can be expensive, and so an essential part of planning is to shop around and find out how to build what you need in a cost effective way. There are many different suppliers of optical components (see appendix C for a non-exhaustive starting point), and shopping around is well worth the time spent. Experienced optical campaigners will have their own opinions on the quality of various components from various suppliers—so asking around your department can also be well worth the time. I will not give my subjective opinions here, but taking the time to fully read the specifications of what you're buying and comparing between suppliers can be revealing!

When budgeting a beamline, it is natural to first think of the 'big' expenses. Things like non-linear crystals, diffraction gratings, finely controllable translation stages or rotation mounts, spectrometers, and cameras are understandably going to be the biggest single items in the beamline budget. However, it is important to not underestimate the cost of simpler components that you may require large numbers of. For example, a simple dielectric mirror may cost only £100, but if you need to buy 25 of them for all the required steering (a not unrealistic number), then they quickly become expensive. Similarly, the cost of relatively simple optomechanics (such as posts and holders) can quickly add up. Thorough planning of the beamline will help you to predict these costs, so you are not taken by surprise at the end. As a ballpark figure, a beamline that the author created for two-colour UV photo-chemistry cost around £16 000, not including the cost of laser systems or OPAs. More complex systems could cost substantially more than this.

8.3 Optical building

Now we come to the question of *how* to actually go about building and aligning an optical beamline. The best way to learn to become a proficient optical builder is simply through practice, but there are general principles worth codifying, and common pitfalls worth avoiding.

8.3.1 Beam steering

Laser beams travel in straight lines, and it is our job to move those straight lines so that they shine on the area we desire. We also know that we need to keep our beams travelling in a plane that is parallel to the surface of the table. If we don't do this, it is not only unsafe, but also a lot of optical elements function best when the beam hits them 'square'. That is, when the beam is collinear with a surface normal coming from the surface of the optical element. The most obvious example of this is a lens, and you can see this easily by looking through your glasses (if you wear them) at various angles and you will see that the lens behaves very oddly when you are not looking into the middle of it. In addition to keeping the beams in the same plane, we often want to direct them along a specific vector within that plane. Mathematically, to define a vector (straight line) we need two points, and so if we are able to align our beam such that it hits two points within our plane, then we have 'aligned' our beam to those points. These points could be irises placed at the entrance and exit of a sample cell, and it is our job to make sure we steer the beam accurately such that it hits both the entrance and exit centrally, and is therefore travelling straight[2]. To accomplish this, we use mirrors to steer our beam through both irises. In general, we will need two mirrors to steer the beam successfully through both irises. The reason for this is illustrated in figure 8.2. A single mirror has to be perfectly placed in order to accomplish this job, and humans are generally not this accurate in placing optics. Having two mirrors will allow the inevitable slight deviations to be compensated for.

Panels (C) and (D) in figure 8.2 show how a beam can be aligned to pass through two irises using two mirrors. In practical terms, to get the beam path perfect, the alignment on the first iris (the near field) is optimised by adjusting the rearmost steering mirror, and then the alignment on the second iris (the far field) is optimised by adjusting the front steering mirror. However, moving the beam with the front mirror will slightly alter the alignment on the first iris, and so this process is repeated

---

[2] Many pieces of equipment, such as OPAs, autocorrelators, or spectrometers, contain an iris at both the entrance port and further inside so that the incoming beam can be reliably steered in straight.

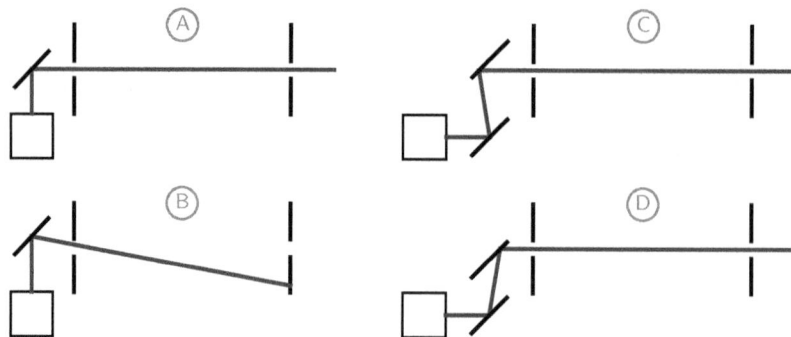

Figure 8.2. Illustration showing the utility of using two mirrors for steering a beam (red) from a laser (black box) through two irises (black). In case (A), the single mirror is aligned exactly with the iris, and no second mirror is necessary. However, slight deviation in the position of this mirror (B) means that coupling into both irises becomes impossible via rotation of this mirror. Using two mirrors, as in (C) and (D), allows slight deviations to be compensated for, as the position that the beam hits the second mirror can be adjusted using the first mirror, and then the angle of the second mirror can be finely adjusted to get the beam travelling through both irises.

in an iterative fashion until the beam passes through both irises perfectly. This process is often referred to as doing a **beam walk**, and is an essential skill to have to be an effective optical builder. Note that the targets need not be irises, but could be marked alignment targets on beam blocks, experimental samples, or even marks on a piece of laser shielding. The fundamental process of using two mirrors to walk the beam and ensure it is travelling along the desired path is the same. Some good rules of thumb when doing a beam walk are:

- Placing the first iris as near to the final mirror as possible, and the second as far from the final mirror as possible, will minimise the number of iterations you need to do.
- Do not be afraid to 'overshoot' with one of the mirrors at first. This will normally mean you converge on the good alignment faster, and you will quickly develop a feel for how the system responds to your input.
- Start with the adjustment screws on each mirror mount set as neutrally as possible, and then get the alignment in the irises as close as you can to perfect by just moving and rotating the mirrors without touching the screws. This will make the final adjustment using the screws quicker, and will ensure that only small adjustments are needed to get the alignment perfect. Running out of screw thread when you are close to perfect alignment is a frustrating experience!

We can also use irises to create targets around beams that we already know are aligned well. A typical example of this might be if you have painstakingly aligned a beam such that it passes through and hits your experimental sample exactly in the right position. Once you have the beam aligned well like this, it is normally prescient to place irises centred around the beam, so that you can easily get the beam aligned back on to your sample simply by following the beam walk procedure described above.

Transmissive optics

When placing a transmissive optic such as a lens into a laser beam, it is normally important that the incoming beam is collinear with a surface normal from the optic. To ensure this, the optic needs to be placed such that it is totally 'flat' with respect to the incoming beam direction. This is difficult to achieve simply by eye, but a simple trick that can help is to look for the **back reflection** from the optic. This is the reflection coming from the face of the optic that the beam hits first, and if the optic is correctly placed, then this back reflection travels back along the incoming beam direction. Whilst this reflection is normally very weak (especially if using an AR coated optic), it can normally be seen faintly if the room lights are switched off, or by using an IR viewer if you have an IR beam. Rotating the optic in its mount will move this reflection, and so it can be exactly lined up with the incoming beam—ensuring that the optic is perfectly placed.

Safe optical building

Building a beamline necessarily requires that there are exposed and unshielded laser beams flying around the table, at least temporarily. Some simple safety precautions can minimise any risk, and also make the process a lot less nerve-wracking. Simple things like attenuating the beam as much as possible before you start to align it, never letting a beam fly into empty space without knowing where it is going, and always taking care to catch unwanted reflections or parts of the beam that are dumped will help to keep you and your co-workers safe. It is important to remember that every glass surface will create a reflection[3], and this includes the front and backside of optics. Make sure that you account for all of these reflections. Another area of potential danger is when using an optic with a clear or polished backside. Beams will fly straight through the clear backsides, and this can create a major safety issue if used in periscopes where vertical beams will occur. Gaffa taping some cardboard to the back of the mirror mount to cover up the backside of the mirror is an easy, quick, cheap, and safe solution.

Another aspect of safety is keeping your expensive optics safe. Very intense ultrashort pulse laser beams can easily damage optics. **Keeping optics clean** is the best way to minimise damage. Most of the time, a burned optic gets burned because some dirt on the surface absorbs laser light and scorches the surface. Keeping dust off of optical setups by covering them up is a good first step, and periodically checking to make sure no optics are damaged is a good habit to get into. If you do find a dirty optic, the first thing to do is to blow some air on it from a clean source to try and remove the dirt. A good source of clean air is an empty and dry solvent squeeze-bottle, as in the author's experience canisters of 'clean' compressed air still contain trace amounts of things you don't want on your optics. If blowing air onto the optic did not remove the dirt, then cleaning it carefully with methanol and lens tissue should remove most of the dirt. However, be certain that the optic and its

---

[3] If you see two intense reflections from a mirror, then it is likely that you have it in back-to-front, and the uncoated backside now is at the front and reflects a lot of light, while the HR-coated front side is at the back and also reflects a lot of light at a slightly different angle.

coating can tolerate methanol (or whichever solvent is used) before you do this—some optics cannot be cleaned in this way (this is especially common with things like wire-grid polarisers).

The other main cause of damaged optics in ultrafast systems is the laser pulses being intense enough to cause non-linear effects like self-focussing. These can lead to the creation of very high local intensities which damage even clean optics. Normally you will see an optic start to glow, or see a bright spot in it, before it is irreparably damaged, so keeping an eye on the beamline when you first allow a high power beam to enter it is a good idea. Self-focussing can be particularly pernicious, as you may inadvertently end up focussing your beam to a point somewhere you're not looking. In the worst case, this could drill a hole in a vacuum window—so it is worth being aware of this if you are using high power beams.

8.3.2 Overlapping beams

The final thing we will discuss in terms of optical construction is how to overlap two beams. When we discuss overlapping beams, there are two things we can mean: the **spatial overlap**, which is ensuring that the beams overlap at the same point in space; and the **temporal overlap**, which is ensuring that they arrive at the same point in time. With ultrafast lasers, finding the temporal overlap can be particularly challenging as to do this we need to make two pulses arrive at the same time as each other to within a few femtoseconds. We will address each of these problems in turn. We assume that both beams originate from the same driving laser—otherwise finding temporal overlap is essentially impossible.

Spatial overlap
How difficult it is to spatially overlap two beams depends mostly on how big the overlapped beam waists are (bigger is easier), and on whether the two beams are initially collinear or not. If you are using large, unfocussed beams, then overlapping two beams in space can be done by eye, simply by holding a piece of card where you want the overlap and steering both beams on top of each other. You may require some fluorescent paper, or IR paper, to see the beams if you are using pulses that are not visible immediately by eye. Overlapping two focussed beams is more challenging, however—especially if the beams are of different colours. For the purposes of the following, we will imagine we are trying to overlap an IR beam with a UV beam.

We consider initially the case where the two beams are collinear. To make these beams collinear requires that they are recombined on a dichroic mirror. Figure 8.3 shows how this is achieved. The IR beam is sent through the back of a mirror where the front side is coated to reflect the UV beam. Special mirrors to do this are available to buy and are often marketed as 'harmonic separators'. Once the beams are collinear, they are then focussed with a lens, and we need to find a way to ensure that the foci both overlap. The immediate question is then to ask how we can actually see *where* the foci are—and we normally do this using something like a beam profiling camera, or photodiode and pinhole, as discussed in the context of beam characterisation in chapter 6. It is then a case of directing the focussing beams onto the camera/diode

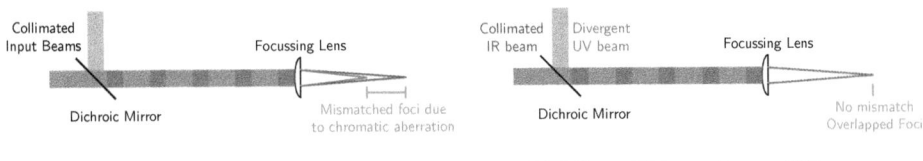

(a) Collimated input beams.   (b) Collimated IR beam, divergent UV beam.

Figure 8.3. Spatial overlap of two different coloured beams using a singlet lens. (a) When the beams are both collimated, chromatic aberration causes a mismatch in the positions of the foci. (b) Making the UV beam slightly divergent (using a telescope) can compensate for the chromatic aberration, and allow the foci to be overlapped in space.

(with attenuation!), then finding one beam, marking its position, and then steering the other beam on top of it without moving the first beam.

This sounds straightforward, but there is one further complication when focussing beams of different colours using a lens: the chromatic aberration of the lens means that the different colours will be focussed to slightly different points along the propagation direction of the laser beams, as illustrated in figure 8.3(a). To compensate for this, we need to adjust the divergence of one of the beams such that the focus of that beam overlaps with the other (recalling that lens focal lengths are quoted for collimated input beams, and so a diverging input beam will have a longer focal length). This is normally controlled by adjusting a telescope placed in one of the beams[4], as illustrated in figure 8.3(b). If the two beams are not collinear, then the process is much the same, except that it can be **vastly** harder to find the overlap position, as collinear beams are already reasonably well overlapped in space by definition. Often in this case, overlapping the beams **using your experimental signal** is the best approach. If you have a clear experimental response from when the beams are overlapped in space, then using this to optimise your overlap is probably the easiest way to go about things, but it is not always the case, or the signal is too small to be easily usable in this way.

Temporal overlap
Overlapping two femtosecond laser pulses in time can be quite challenging. Light will travel about 10 µm in 35 fs, and so you need to be able to adjust the lengths of each beam path with micrometre precision. This will require that at least one of your beams travels on a delay stage controlled with a micrometre screw, so you can accurately move your beams in such small increments. The next thing to do is to get a rough estimate for how far apart your pulses will be in time. This can be done either by using an oscilloscope and photodiode, or by simply measuring both beam paths with string and a tape measure. Both of these methods are accurate to within around a nanosecond or so—and a useful rule of thumb is that light travels around 30 cm in a nanosecond. This will tell you if you need to add vast amounts of delay to one of the beams—as may be the case if one has passed through an OPA and one has not.

---

[4] The problem could also be entirely circumvented by using an **achromat**, but these are relatively thick and so not ideal for ultrafast use.

Having to add several metres of optical delay to a line to synchronise it with a line that has gone through an OPA is not uncommon. Getting them synchronised to within 30 cm is about as far as electronic assistance can take us, and now we need to use an optical method to determine if the pulses are temporally overlapped.

We already know from chapter 6 that non-linear crystals can be used to measure ultrashort pulse durations, and when the pulses are overlapped in the crystal both spatially and temporally a non-linear interaction occurs which produces a new colour of light we can detect. This is one of the most common ways to find the point in time where two ultrashort pulses are overlapped—this point in time is colloquially known as t_0 ('t-zero'). What crystal we need depends on what non-linear interaction we are looking for, but for illustrative purposes we will assume we are looking for the temporal overlap between an 800 nm IR beam and a 266 nm UV beam.

The first task is to find out what non-linear interaction we will need to look for. In the example here, we can do a difference-frequency mixing of 800 nm and 266 nm to produce 400 nm light. This requires that we have a non-linear crystal that can phase-match this process, and a useful tool to find out what we can use is at toolbox.lightcon.com, as described in detail in chapter 4. In this case, we find that a BBO crystal cut at 44.3° is what we need. The beams need to be spatially overlapped in this crystal, and a simple way to do this (provided that your beams are intense enough) is simply to place the crystal in the beamline before the focussing lens, where the two beams are already collinear. This is illustrated in figure 8.4.

Having placed the crystal here, we should be able to make 400 nm light when the pulses are overlapped in time. To be able to actually see this light, it is useful to disperse the collinear output using a prism, and then looking at the different colours on a white card (or similar). Then by scanning the delay stage, you should be able to see a very brief flash of blue 400 nm light when the pulses are overlapped in time. When you see this, record the position of the delay stage, as this is your t_0! Note however, that the t_0 you record here may not correspond exactly to the t_0 in your experiment, as things like vacuum windows and focussing lenses will shift the

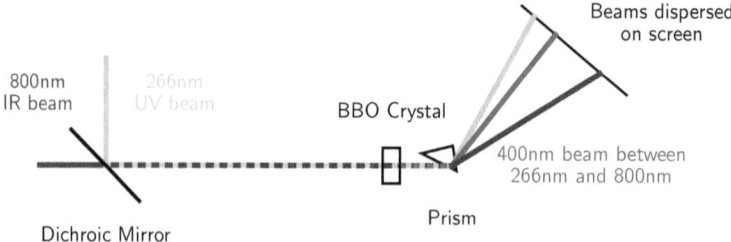

Figure 8.4. Illustration of a setup used to find temporal overlap between an 800 nm IR beam and 266 nm UV beam. The beams are spatially overlapped in a BBO crystal, and the beams are then dispersed onto a screen using a prism. The delay between the IR and UV pulses is scanned, and when the beams are overlapped temporally, a 400 nm beam will appear on the screen.

temporal overlap somewhat (normally by a millimetre or less). You can add a 'fake window' before your crystal to account for this if desired, or place a mirror between the focussing lens and window (if space permits), and send the reflected beams into a crystal to account for the effect of the lens as well.

The process described above sounds straightforward, but in reality trying to find t_0 can be a very frustrating experience if you are not lucky enough to find it in the first few attempts. If you do not see it, then there are a few things you can try to do:

- **Optimising the crystal angle**: As the phase-matching is exquisitely sensitive to the angle of the non-linear crystal, iterating a process of adjusting the angle slightly, then scanning the delay stage, may help you to find the overlap if the crystal angle is not perfect when it is placed flat in the beam path.
- **Scanning more slowly**: While watching a piece of card can be tedious, the temptation to scan the stage very quickly can be problematic. Remember that you may only have a few tens of microns of stage movement where you will see the t_0 light produced.
- **Increasing the intensity**: Non-linear processes depend heavily on having intense light. Sometimes unfocussed beams will not have sufficient intensity to drive the non-linear process, so focussing the beams into the crystal can help. Take care not to drill a hole in the crystal!
- **Checking polarisations**: Ensuring that you have the right combination of polarisations for your desired phase-matching scheme is important.
- **Optimising the spatial overlap**: Especially if you are focussing the beams, it is critical that the beams are well overlapped within the crystal.
- **Using a photodiode**: A photodiode is more sensitive than your eyes, so using a diode with a filter attached to it that will only let through the light from your non-linear process can help you see the t_0 if it is very faint. A spectrometer can also be useful here.
- **Using an experimental signal**: If all else fails, and you already have some experimental signal you can use, then looking for the temporal overlap within the experimental signal can be a good option. Especially if the experimental signal behaves more like a step function, which turns 'on' when the beams overlap and then stays at a high level after the overlap for some time. This can make it much easier to find the overlap, as you are not looking for a very transient flash of light.

Once you have some faint t_0 signal, then often you can make it much more intense by optimising things like the overlap of the beams and the angle of the crystal. The good news is that once you have it, it will be easy to find again, and you are well on the way to having a fully-functioning beamline. Finding t_0 is the sort of event that generally merits a celebratory coffee-and-cake break in most labs.

To tie together all of the previous chapters, we will now complete our story in the next chapter with a worked example. We will walk through the design and construction of a beamline that was constructed by the author to UV-pump, IR-probe gas-phase photochemistry.

Chapter 9

Case study: pump-probe beamline

In this final chapter we will tie together the practical knowledge discussed in the preceding chapters by walking through the design of a beamline from start to finish. The beamline in question is one that was designed and built by the author to do ultrafast pump-probe spectroscopy on gas-phase molecules. The beamline layout is shown on a full page in figure 9.1, and we will refer to it periodically throughout this chapter.

9.1 Initial equipment

As in many projects, there was not a blank sheet of paper (and blank cheque) available, and the beamline had to fit within existing constraints. Principally, these consisted of the driving laser system and OPA, which already existed prior to the beamline design. The optical tables were reasonably empty, so there were no huge spatial constraints. The budget was relatively tight as it was a new research group starting from scratch, so many other kinds of equipment also had to be bought.

The beamline is driven by a Spectra-Physics Solstice Ti:Sa laser system. This produces 800 nm, 35 fs, 5 mJ pulses at a repetition rate of 1 kHz. 3.4 mJ of this output is steered into a TOPAS Prime OPA, which is coupled to a NirUVis mixer extension box. This OPA and mixer give a tunable output from 190 nm to 2600 nm, whilst maintaining approximately the pulse duration of the driving laser. The pulse energy of this beam varies substantially with wavelength, but in the UV region we are interested in, it is always greater than around 10 μJ. The height of the output of the Ti:Sa laser is around 160 mm from the table surface, but the height of the OPA and mixer output is 130 mm from the table surface. The beam from the Ti:Sa output is around 11 mm in diameter, but the beam from the OPA is substantially smaller, around 5 mm in diameter.

9.2 Requirements

This beamline is intended for use in a gas-phase UV-IR pump-probe spectroscopy experiment. The UV pulses will be used to initiate ('pump') some photochemistry in an isolated molecular target in vacuum, and then the IR pulses will Coulomb explode

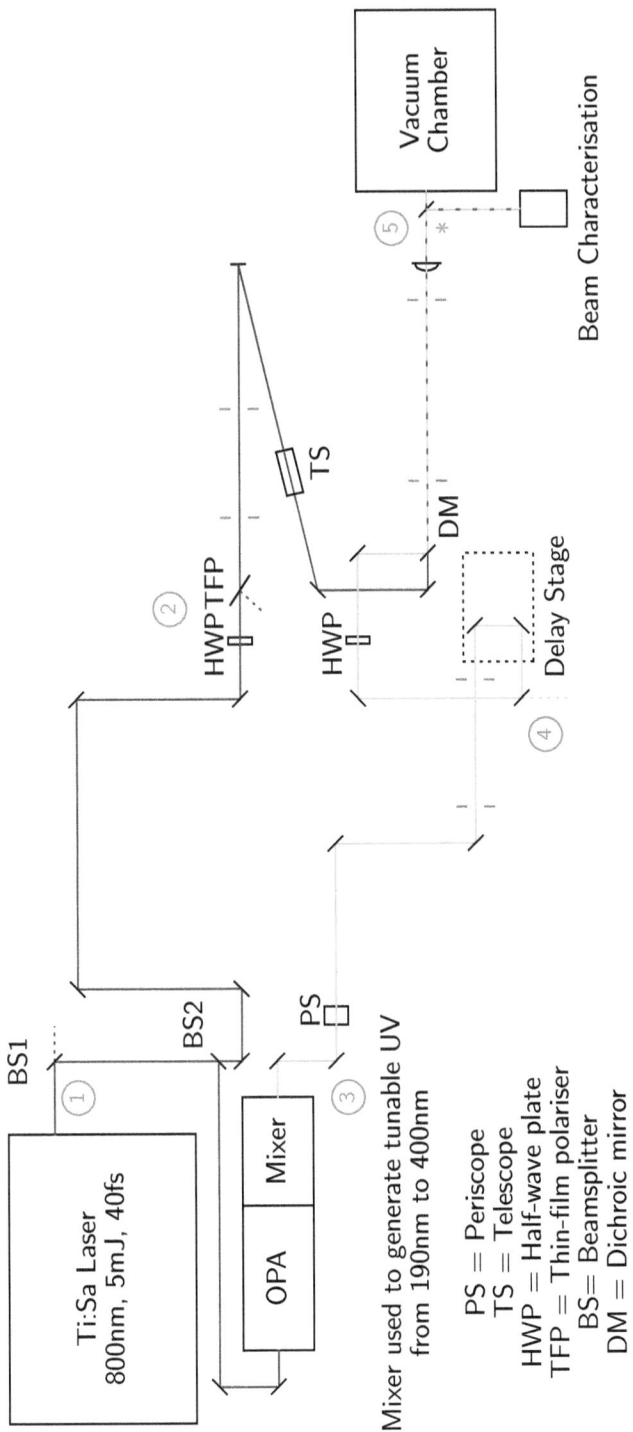

Figure 9.1. A UV-IR pump-probe beamline. See text for details.

('probe') the target. Coulomb explosion is a process where a molecule is subjected to a very intense laser field and undergoes rapid multiple ionisation, which causes the molecule to 'explode' into ion fragments almost instantaneously. Knowing these requirements, we can create a list of the laser conditions we need to have in our vacuum chamber.

- **Beams:** We need to overlap two beams in space and time in our vacuum chamber, and delay them relative to each other with femtosecond time precision.
- **Position:** The target is a molecular beam inside the vacuum chamber, and the window to the chamber is around 160 mm higher than the surface of the optical table.
- **Colour:** The IR pulses can be used at 800 nm, but the UV pulses need to be tunable between around 200 nm and 400 nm.
- **Size:** The IR pulses to induce Coulomb explosion need to be fairly intense, so a small 25 μm beam waist is desirable. The UV pulses can be somewhat larger, as the same intensity is not required, and having a larger beam waist means that spatial overlap is easier, and we ensure that we are only probing molecules that are actually pumped. Around 50 μm is more than sufficient for the UV beam.
- **Pulse Duration:** Measurement of sub-picosecond dynamics is desirable, so the pulses should not be longer than around 100 fs. Adding additional compression may also stretch the budget somewhat, however.
- **Intensity:** Around 1×10^{14} W cm^{-2} is required for Coulomb explosion. Several orders of magnitude less is required for the UV pump, around 1×10^{10} W cm^{-2}. This translates into around 100 μJ of pulse energy for the IR pulses, and less than 1 μJ for the UV pulses.
- **Polarisation:** Variable linear polarisation of both beams is required.

We now compare these requirements to what our driving laser can provide. Working down the above list, we find the following:

- **Beams:** We will need to add a motorised delay stage to either the pump or the probe arm, to allow the temporal overlap to be found.
- **Position:** We will need to raise the UV beam by 30 mm up to a height of 160 mm.
- **Colour:** The OPA provides all of the needed tunability.
- **Size:** We will need to focus the beams into the vacuum chamber. The distance between the edge of the laser table and the position of the molecular beam is around 20 cm, so using a 300 mm focal length lens for focussing would be ideal[1]. Using this lens with our direct 11 mm IR output beam would produce a beam waist of around 12.5 μm, so we would like to *shrink* the IR beam by a factor of two. The UV beam may also be too small, but adding a UV telescope may make the pulses unacceptably long, so we will see how it looks once we have beams in the chamber, as the OPA output may be quite divergent anyway.

---

[1] So the lens is not clamped right on the edge of the table.

- **Pulse Duration:** We should plan the beamline and then calculate the GDD added by all of the transmissive optics, and the air travelled through.
- **Intensity:** These pulse energies are easily obtainable using our laser system.
- **Polarisation:** Adding half-wave plates to both beam lines will be necessary.

This beamline will also be used to drive two other experiments with more or less identical requirements to this (though not simultaneously), so we should also keep this in mind when planning. It was agreed with the other groups using the beamline that we would just provide a place where beams could be easily 'picked off' and sent to another experiment, but that they would build their own beamline. Areas where a beam has been picked off and sent to another experiment are shown as dashed red and blue lines that trail off into space on figure 9.1. We now know what requirements we have to keep in mind whilst planning the rest of our beamline.

9.2.1 Constraints

As shown in figure 9.1, the vacuum chamber is not directly adjacent to the output of the laser system, and there is around 3 m of optical table in between the laser output and vacuum window. We also need to be able to provide beams to drive other experiments, so we must account for this. Budgetary constraints are also such that expensive items (like the motorised delay stage) should ideally be able to be used by multiple experiments. Adding additional pulse compression stages may also be difficult, so we should keep an eye on the accumulated GDD.

9.3 Design and construction

We will now walk through figure 9.1, discussing how all the requirements will be met and any special considerations that need to be made. We will make reference to the orange circled numbers as key reference points throughout our discussion.

Starting at point (1), with the direct Ti:Sa output. This output needs to be split up such that it can drive the OPA, and also so that a small fraction can be sent into another experiment (the dashed line after the first beamsplitter). To accomplish this, two beamsplitters are placed at the direct output. Beamsplitter BS1 is a 90:10 beamsplitter, and so reflects 90% of the incident light whilst transmitting 10%. This 10% (500 μJ) is sent to another experiment. The remaining 4.5 mJ is directed onto BS2, which is an 80:20 beamsplitter. The reflected 80% (3.6 mJ) is used to drive the OPA, being reflected off two other 1.5 inch mirrors to steer it into the OPA. The transmitted 900 μJ will become the probe beam for our experiment, and for one other experiment. Both of these beamsplitters are 1.5 inch beamsplitters provided by Spectra-Physics with the laser. This beam is sent on a circuitous route to delay it relative to the UV beam, as the UV beam is delayed by around two metres having passed through the OPA. All the mirrors used in the 800 nm IR line are low-GDD broadband dielectric mirrors designed specifically for ultrafast pulses of this wavelength, and are all 1 inch in diameter.

We said that for our experiment that we wanted to shrink the size of the IR beam down, and so need to use a telescope. We also know that generally it is good to shrink the beam earlier rather than later, so it is not difficult to align and any divergence can be controlled. In this case, however, the two experiments being driven by this beam have quite different requirements for the beam size, so we delay shrinking the beam in a telescope until after we have split this beam further. This further splitting happens at point (2), with the typical half-wave plate/polariser combination, which also acts as a variable attenuator. The beam that is transmitted through this polariser will become our probe beam, whereas the reflected beam will become the probe beam for another experiment. As both experiments will not run simultaneously, this can also be used as a variable attenuator for both experiments. The attenuator consists of a zero-order half-wave plate, and a broadband thin film polariser. The polariser is 2 inches in diameter, as it needs to be placed at Brewster's angle, so the beam is spread out over the surface of it, and would clip on a 1 inch polariser. After this attenuator, we have the first set of irises that can be used to aid alignment.

The IR probe beam then reflects off a zero-degree mirror and passes through a telescope. This telescope was chosen to shrink the beam down by a factor of two, so consists of a converging lens with +200 mm focal length at the input, and a diverging lens with −100 mm focal length at the output. Both of these lenses are AR coated for 800 nm to minimise power loss. These lenses are placed 100 mm apart (the sum of the focal lengths), and the beam is shrunk by a factor of two (the ratio of the focal lengths). The beam is then sent steered through the back of the dichroic mirror DM, and into the vacuum chamber. The dichroic mirror DM has the front face coated to approximately reflect 250 nm to 270 nm, and the back face AR coated at 800 nm to minimise transmission losses. Note that two mirrors are used to steer this into the chamber after the telescope, allowing easy alignment on the two irises placed before the lens into the vacuum chamber window at (5).

The UV beam is produced at the point (3) from the OPA and mixer, and is immediately raised using a periscope up to 160 mm, such that all the beams are propagating in a plane 160 mm off the table, at the same height as the vacuum window. It was chosen to do this rather than send the beam down to the table (for increased stability) and then bringing it back up, as avoiding periscopes was desirable, and the entire table is shielded from external air currents so stability was not a major concern. As the UV beam is tunable, it is important that mirrors reflect a broad range of UV bandwidth, so that they don't have to be replaced frequently when moving the UV wavelength around. For this purpose, protected aluminium mirrors were judged to be the best solution, as they have a high reflectivity down to around 250 nm, and reflect a broad bandwidth. All mirrors not specifically labelled in the UV beamline are protected aluminium.

The UV beam then is directed through some irises onto the delay stage at point (4). Irises are placed before the delay stage, as it is imperative that the light coupling into the stage is travelling exactly straight. For this reason the irises were screwed straight into the bolt holes of the laser table, so that the light was always travelling directly down the bolt holes. Substantial care was then taken to ensure that the angle between the two mirrors on the stage was exactly 90°, so that the beam coming off the stage was also travelling exactly straight. This is imperative, as if it is not, then the pointing direction of the UV beam will change whenever the stage is moved—spoiling the spatial overlap in the vacuum chamber. The best way to check this is to look at the beam coming off the stage as far away from the stage as is reasonable, and ensure that the beam does not move at all when the stage is driven.

Coming off the delay stage, the beam is then directed onto a mirror that can be rotated to send the UV beam either towards our vacuum chamber here, or the other experiment using the UV beam (dashed blue line). The beam is then sent through a half-wave plate that is designed for 266 nm in the first instance, but which will also be effective at wavelengths near this wavelength. It is then steered onto the dichroic mirror DM, where it is reflected through the irises into the vacuum chamber at (5). The UV beam was deliberately reflected off this mirror (rather than sent through it) to minimise added GDD.

Both beams are directed through a focussing lens at point (5), and this lens is a 300 mm singlet lens made of UV fused silica which is AR coated at 800 nm. Singlet lenses are preferable as they are thinner, and so add less GDD—but chromatic aberration needs to be compensated for as we need to overlap the 800 nm beam with the 266 nm beam. In this beamline, we would adjust the divergence of the IR beam with telescope TS to overlap the foci on the molecular beam. The lens is also not AR coated for the UV wavelengths, as the lens would have to be custom coated for this, which would be prohibitively expensive. This means that there will be some UV power lost to reflection (around 5%), and also an additional stray beam that needs to be safely blocked. The vacuum window the beams are focussed through is also made of UV fused silica, and is AR coated for 800 nm too. The window is 4 mm thick.

A final note about point (5), where there is a mirror shown by an orange asterisk. This mirror is removable, and is placed after the focussing lens to direct the focussed beams down towards the beam characterisation equipment. In practice, this is where you would put a beam profiling camera or a photodiode to look at your focal spot sizes, and to make sure the beams are overlapped. Doing it as close as possible to the vacuum window ensures that when the mirror is removed, the beams remain overlapped in the vacuum chamber.

9.3.1 Pulse durations

We can now see that this beamline will perform as we expect, and we have chosen all the optical elements such that it will perform to the requirements we set initially. However, what we have not yet done is consider the final pulse durations in our chamber. We know the pulse duration at our laser output, and we can estimate the final pulse duration in our chamber by calculating how much GDD will be added to each beam as it propagates. We will do this for each beam in turn.

In all cases, the numbers for GDD were either obtained from the manufacturers of different optics, or (in the case of transmissive optics) found in online databases of GDD data of certain materials, where the GDD added by a given thickness of a material can be easily looked up.

IR beam
Ideally the IR output is a transform-limited 35 fs 800 nm pulse. We then need to calculate the GDD added to this from each optical element. This is easily done using

online calculators, such as the light conversion optical toolbox, and the results are summarised in table 9.1. 'Worst-case' numbers for the GDD have been assumed throughout, so this gives an upper estimate for the final pulse duration.

So, in a worst-case scenario, our initially transform-limited IR pulse will end up broadened by almost a factor of 3 by the time it gets to our vacuum window! We can compensate for this to some extent by moving the compressor inside the laser system, but we cannot move this too far as the OPA is very sensitive to input pulse duration, and moving it a long way will severely impair the function of the OPA. This number may seem alarming, but it is likely to be a gross overestimate, as the numbers assumed are all worst-case numbers, and if the laser output is not fully transform-limited, then the broadening will also be less severe. A pulse duration of around 70 fs in the vacuum chamber is likely to be a more realistic estimate.

UV beam

We now perform the same analysis with the UV beam. Here we do not know the exact output pulse duration, but it is likely to be slightly broader than the IR pulse duration after all of the processes occurring in the OPA. For the purpose of this calculation we will assume we start with a 45 fs, 266 nm pulse. The results are shown in table 9.2.

Table 9.1. Effect of accumulated GDD on IR pulse duration.

Optical element	Quantity	GDD per element (fs^2)	Cumulative pulse duration (fs)
800 nm mirrors	9	30	41
HWP	2	80	49
TFP	1	120	56
Telescope	1	160	67
Focussing lens	1	80	72
Air	5 m	100	79
Vacuum window	1	160	91

Table 9.2. Effect of accumulated GDD on UV pulse duration.

Optical element	Quantity	GDD per element (fs^2)	Cumulative pulse duration (fs)
Al mirrors	11	20	47
HWP	1	400	59
Dichroic mirror	1	30	60
Focussing lens	1	400	79
Vacuum window	1	800	123
Air	3 m	300	140

As we can see, the UV pulses are substantially more affected by the broadening. It should be noted that again these are 'worst-case' numbers, as the Al mirrors probably add much nearer to zero GDD than this number, which was taken as the manufacturer's quoted upper limit. However, this illustrates that air and transmissive optics have the potential to cause lots of unwanted broadening for these pulses. An additional UV compressor may be required if much shorter UV pulses are needed, but the numbers here are considered to be reasonable as a place to start.

IOP Publishing

Ultrafast Lasers and Optics for Experimentalists

James David Pickering

Appendix A

Electromagnetic waves

In my experience people entering this field from a non-physics background (such as chemistry or biology) can struggle with some of the concepts surrounding the mathematical description of electromagnetic waves. For a trained physicist, this is generally just assumed knowledge and so it can be demoralising for a new student to not understand what is meant by terms like 'phase', or 'k-vector'. However, it is rather straightforward if broken down simply. A more mathematical treatment can be found in a standard text on electrodynamics, such as reference [1].

A one-dimensional travelling electromagnetic wave $E(z, t)$ can be expressed as follows:

$$E(z, t) = E_0 \exp[i(kz - \omega t)] \qquad (A.1)$$

Physically, this wave could be the electric field of some laser light. The wave is 'travelling' because it is moving in both space (the coordinate z), and time t. Let us consider the meaning of each of the terms in equation (A.1) in turn.

- E_0 refers to the **amplitude** of the electric field—the maximum height of the peaks in the wave. This would have units of volts per unit length.
- t is time, with units of time.
- ω is the **angular frequency** at which the wave oscillates *in time*. It has units of radians per unit time, such that the product ωt has units of radians, which are dimensionless[1]. This is normally just called the **frequency**. *The more cycles the wave completes per unit time, the larger the frequency.*
- z is the position of the wave in space along the z-axis, with units of length.
- k is the **angular wavenumber** of the wave. This has units of radians per unit length, such that the product kz has units of (dimensionless) radians. It is normally just known as the **wavenumber**. This can be thought of as a **spatial frequency**, where ω was a **temporal frequency**. *The more cycles the wave completes per unit length, the larger the wavenumber.*

---

[1] A radian is defined as the ratio of arc length to radius length of a circle, thus the units of length cancel out and the radian is dimensionless.

- i is the imaginary unit, defined such that $i^2 = -1$. This will be discussed further below.

The wave is written in an exponential form, but is really just a sinusoidal wave, as we know from Euler's formula that:

$$\exp(i\theta) = \cos(\theta) + i\sin(\theta) \tag{A.2}$$

So if we took the real part of our wave $E(z, t)$:

$$\text{Re}[E(z, t)] = E_0 \cos(kz - \omega t), \tag{A.3}$$

which is the sinusoidal form we expect. We use this exponential form as it is the most general, and it makes manipulating the wave much simpler when we try to add phase factors and things. To visualise all the parameters above, it is easiest to plot the wave, but before we do this there is some mathematical complexity that needs to be cleared up.

You will probably encounter multiple definitions of the wavenumber, unfortunately. Within spectroscopy and chemistry it is normally thought of as the reciprocal of the vacuum wavelength of a particular spectroscopic transition, with the symbol $\tilde{\nu}$:

$$\tilde{\nu} = \frac{1}{\lambda} \tag{A.4}$$

This definition is useful in chemistry where the desire is really just to have a number that is linked to the transition wavelength but is directly proportional to transition energy. However, in the context of laser physics, we define the wavenumber, k, as:

$$k = \frac{2\pi}{\lambda}, \tag{A.5}$$

which has units of radians per unit length, as mentioned above. This is really the 'angular wavenumber', but it is normally just called the 'wavenumber', like the angular frequency is just called the frequency. We do this because we are always talking about waves that are periodic, and what is interesting is how often the wave completes a complete periodic 'revolution' around 2π radians (a circle). So how many radians our oscillating wave moves through in a propagation length, or propagation time, is what interests us.

With this in mind, we should think about how we can link together *the number of oscillations per unit length* (k) with *the number of oscillations per unit time* (ω). It seems natural that these should be connected: if the wave is oscillating through a certain number of radians in a certain time, and is also moving through space, then the **speed at which it moves through space** will dictate how many radians it oscillates through in a given distance. That is:

$$k = \frac{\omega}{v_\text{p}}, \tag{A.6}$$

where v_p is the **phase velocity** of the wave, which is how fast it is moving in whatever medium it is travelling in. If the wave moves faster, v_p is larger, and the wave won't have time to oscillate through as many radians in a given distance than it would if it was moving more slowly. This is simply what equation (A.6) expresses mathematically.

Figure A.1 shows graphically all of the quantities we have discussed, and how they relate to one another. We have also defined the reciprocal of the frequency, the **oscillation period**, $T = 2\pi/\omega$, in the leftmost plot on figure A.1; and the **wavelength**, λ, in the rightmost plot. Both of these quantities represent the time taken (T) for, and the distance travelled (λ) in one full oscillation (through 2π radians). It is clear from the figure that the wavelength, λ, can be defined in terms of previously met quantities as:

$$\lambda = \frac{2\pi}{k} = \frac{2\pi v_p}{\omega} \tag{A.7}$$

However, at the start of this section we specifically said that we would clarify the meaning of **phase** and **wave vector**—two of the concepts that cause most confusion in this topic in my experience of teaching it. Before we discuss the phase in the following section, we will briefly discuss the wave vector, $\boldsymbol{k}$.

A.1 Wave vectors

The wave vector $\boldsymbol{k}$ looks like the wavenumber k, but is in bold. This is because the wave vector is a **vector**, so has a direction and a magnitude. **The magnitude of the wave vector is simply the wavenumber.** The wave vector points in the direction of propagation of our electromagnetic wave (the direction of propagation of a laser beam, for example). In the case that the wave is travelling in 3D space, rather than the 1D case shown above, then the direction of travel can be split into three

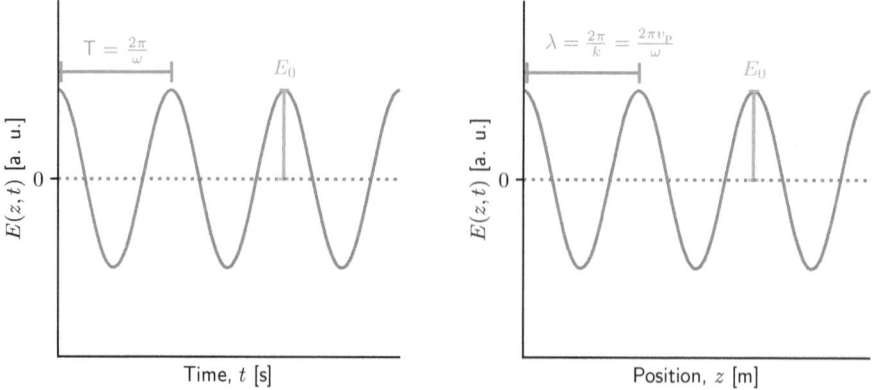

Figure A.1. Illustration of our wave $E(z, t)$ plotted as a function of time (left) and position (right). Quantities discussed in the text are annotated.

components, corresponding to movement along the x, y, and z axes. In this case we make the substitution:

$$kz \rightarrow \boldsymbol{k} \cdot \boldsymbol{r} = k_x \hat{\boldsymbol{x}} + k_y \hat{\boldsymbol{y}} + k_z \hat{\boldsymbol{z}}, \tag{A.8}$$

where $\boldsymbol{r}$ is our position vector in 3D space, which can be split into three components in terms of the unit vectors $\hat{\boldsymbol{x}}$, $\hat{\boldsymbol{y}}$, and $\hat{\boldsymbol{z}}$. The magnitude of each of these components is k_x, k_y, and k_z respectively. The total magnitude k_{3D} of our 3D wave $\boldsymbol{k}$ is given straightforwardly by:

$$|\boldsymbol{k}| = k_{3D} = \sqrt{k_x^2 + k_y^2 + k_z^2} \tag{A.9}$$

The value k_{3D} given in equation (A.9) is simply the three-dimensional analogue of the one-dimensional k used equation (A.1). To summarise, the important points are:

- The wave vector, $\boldsymbol{k}$, is a vector that points in the direction that the wave is propagating in.
- The magnitude of the wave vector is the wavenumber, k, and tells you how many cycles the wave completes in a unit propagation distance (units of m^{-1}).

In the context of lasers, we often talk about **k-vectors**, and these are just another name for wave vectors. Specifically, we talk about them in the context of momentum conservation in phase-matching in non-linear optics. This is because we can link the magnitude of the k-vector k to the momentum of the wave p:

$$p = \hbar k \tag{A.10}$$

A simple dimensional analysis illustrates this. Momentum has units of $\mathrm{kg\ m\ s}^{-1}$. The magnitude of the k-vector has units of $\mathrm{rad\ m}^{-1}$ (as discussed above). The reduced Planck's constant $\hbar$ has units of $\mathrm{kg\ m^2\ s^{-1}\ rad^{-1}}$. So, a larger k-vector (or shorter wavelength, or higher frequency) corresponds to a higher momentum for the wave. We will now turn to a discussion of **the phase**.

A.2 Phase

The concept of phase elicits a lot of confusion among students in my experience, but it need not. Fundamentally, **the phase of a wave tells you which part of the cycle it is in.** For example, a wave with a phase of π rad is halfway through a cycle, and a wave with a phase of 0 rad is at the beginning of a cycle. As such, **the phase is an angle**, given in radians. Recall, however, that a radian is dimensionless, so you will equally see it said that the phase is dimensionless. The phase is generally given between 0 and 2π, but sometimes the phase can be greater than 2π, if a wave finishes a cycle and goes on to the next one[2].

So the phase is just an angle that tells you which part of the oscillation period you are in. If it seems odd that we use an angle to define this, remember that our wave is

---

[2] This lies at the heart of the concept of **phase unwrapping**. If you are looking at the phase of a signal over a long time, then you might see lots of jumps as the phase gets to 2π and then skips back to zero. Unwrapping the phase gets rid of these jumps and gives you a continuous phase.

a periodic sinusoidal function, as shown in equation (A.3). The argument that this function takes is an angle, so it's natural that we use an angle to define where 'on' this function we are. However, we can distinguish between the **absolute phase** of the wave, and a **phase shift** or **accumulated phase** that is added to the wave.

The **absolute phase** tells us exactly where we are in the wave cycle overall. To find the absolute phase of our one-dimensional travelling wave, we note that we wrote equation (A.1) in exponential form deliberately. This exponential form may be familiar from the study of complex numbers in mathematics, where a complex number N can be written as:

$$N = |N|\exp(i\Theta), \tag{A.11}$$

where $|N|$ is the modulus (magnitude) of the complex number, and Θ is the argument (phase) of the complex number. So we can identify the stuff in the exponent next to the imaginary unit as our phase. This means that the absolute phase of the wave in equation (A.1), which we will call Θ, is given by:

$$\Theta = kz - \omega t \tag{A.12}$$

Both quantities kz and ωt are angles as discussed above, so the difference $kz - \omega t$ is also an angle, and defines the absolute phase of our wave. You may sometimes see it written as:

$$\Theta = kz - \omega t + \arg(E_0), \tag{A.13}$$

which accounts for the possibility that the amplitude E_0 is also a complex number with its own phase. The argument of the complex number is just the phase, so $\arg(E_0)$ is the phase of the amplitude which also contributes to the total absolute phase. But in the examples we consider here E_0 is simply a number, so does not have a phase (or has a phase of zero).

We can further exploit the beauty of the exponential form of complex numbers to understand what we mean by a **phase shift** or **accumulated phase**. A phase shift is when we move our wave along in its cycle by a given angle. For illustration, we will call this angle ϕ. Ultimately what we are doing with a phase shift is:

$$\Theta' = \Theta + \phi, \tag{A.14}$$

where Θ' is the absolute phase of the wave after the shift, Θ is the absolute phase of the wave before the shift, and ϕ is the added phase. The exponential form of complex numbers makes this trivially simple. To shift our initial travelling wave $E(z, t)$ by ϕ radians, we simply do:

$$E(z, t) \times \exp(i\phi) = E_0 \exp[i(kz - \omega t + \phi)] \tag{A.15}$$

So multiplication by the **phase factor** $e^{i\phi}$ caused a phase shift of our wave by ϕ radians. This is what we mean when we talk about phase shifts or accumulated phase, here we have accumulated a phase of ϕ radians.

We can get a feel for what this looks like by considering what happens if we shift our wave by some different values of ϕ. This is illustrated in figure A.2. Here we still

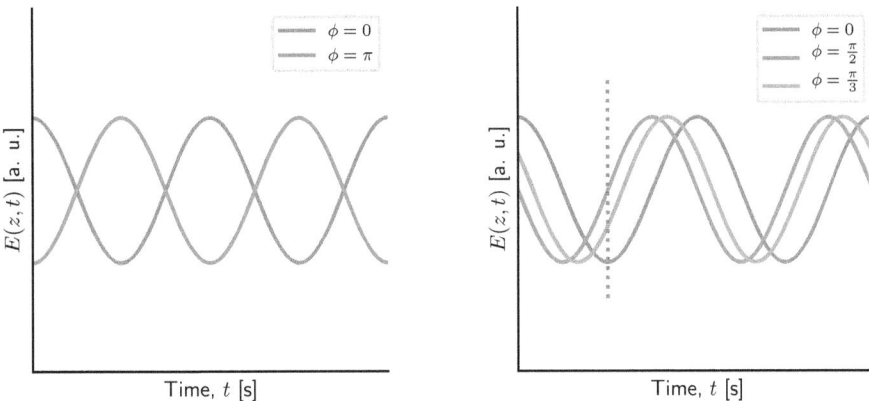

Figure A.2. Illustration of our wave $E(z, t)$ plotted as a function of time with different phase shifts, ϕ, applied.

plot our waves as a function of time, but you can see that the added phase has the effect of moving the part of the cycle that a wave is in at a *given time* around. That is, if you were to pick a specific point on the time axis (such as that shown with the dashed line on the rightmost plot in figure A.2), then all the waves are clearly at different points in their cycle. The orange curve is $\pi/2$ radians ahead of the blue curve, and so on. This kind of plot can be difficult to visualise as it looks as though the waves with a positive phase shift are being moved backwards. The best way to think about it is to look at the unshifted wave (blue), and then look at where it is at the dashed line. Now consider where the blue wave would be in $\pi/2$ radians time—this is where the orange wave is at the dashed line.

To end with a laser based example, often we talk about the spectral phase $\phi(\omega)$ that is accumulated by a pulse as it propagates through a medium. This just means that a wave in the pulse with frequency ω is moved in its cycle by $\phi(\omega)$ on propagation through the medium. $\phi(\omega)$ could be a complicated function of ω, which is what gives rise to the GDD and higher order dispersions that cause our pulse to broaden during the propagation.

Reference

[1] Griffiths D J 2017 *Introduction to Electrodynamics* 4th edn (Cambridge: Cambridge University Press)

IOP Publishing

Ultrafast Lasers and Optics for Experimentalists

James David Pickering

Appendix B

Useful resources

The internet contains a wealth of useful resources that are often much more easily accessible to a working scientist than a traditional textbook, and can present ideas at a more accessible level than many journal articles. Here is a selection of those that are most useful for people working within optics and lasers—all of these have helped me in the past, and many of them have been drawn on when writing this book.

- **RP Photonics Encyclopedia**: www.rp-photonics.com/encyclopedia.html
- **Light Conversion Optics Toolbox**: http://toolbox.lightcon.com
- **Newport Optics Technical Notes**: www.newport.com/resourceListing/technical-notes
- **Thorlabs Technical Resources**: www.thorlabs.com/navigation.cfm?Guide_ID=2400
- **Edmund Optics Technical Notes**: www.edmundoptics.com/knowledge-center/technical-literature/
- **Online Refractive Index Database**: https://refractiveindex.info/

Appendix C

Suppliers

Sometimes, you may not be in a research group with a lot of experience using optics and lasers, so knowing what suppliers exist for various items of equipment can be challenging. Here, a list of commonly-used suppliers of optical components is provided for general reference. This list is intended to be a useful starting point rather than be exhaustive, and a much larger directory of suppliers can be found at www.rp-photonics.com/bg_suppliers.html.

- **Thorlabs**: www.thorlabs.com—very wide product range.
- **Newport Optics**: www.newport.com—very wide product range.
- **Eksma Optics**: www.eksmaoptics.com—wide product range, including non-linear crystals.
- **Layertec**: www.layertec.de—very wide range of coated optics.
- **Edmund Optics**: www.edmundoptics.co.uk—very wide product range.
- **Alphalas**: www.alphalas.com—laser crystals and polarisation optics.
- **4lasers**: www.4lasers.com—wide range of optics and crystals.
- **Crystran**: www.crystran.co.uk—wide range of windows, lenses, and prisms.
- **United Crystals**: www.unitedcrystals.com—non-linear crystals.
- **Lattice Electro Optics**: www.latticeoptics.com—wide range of coated optics.
- **Ultrafast Innovations**: www.ultrafast-innovations.com—wide range of chirped mirrors.

CPSIA information can be obtained
at www.ICGtesting.com
Printed in the USA
BVHW011401060222
628173BV00003B/23